The Chrysalis Teachings
Pathways to the Future

CATACLYSMS??
A NEW Look At Earth Changes

Norma Green Hickox

CATACLYSMS??

A *NEW* Look At Earth Changes

A Blue Star Productions Publication

ISBN 1-881542-28-9

First Blue Star Printing: March 1997

Published by Blue Star Productions
A Division of Book World, Inc.
9666 E. Riggs Rd. #194, Sun Lakes AZ 85248
Ph: 602.895.7995 Fax: 602.895.6991

Printed in the United States of America

Acknowledgments

I wish to express my gratitude to my husband Lyle and my son Neil, both of whom have helped immensely with the Chrysalis Teachings, each in their own way.

I'd especially like to thank my son Neil for taking responsibility for doing the computer work on the graphics in the books. I'd also like to acknowledge his expertise at not only doing the graphics but also all the other computer problems that came up. Without him I could not have kept my equipment operating.

I want to thank Michael Bissonnette for his help, encouragement and support. Without his insightful and intelligent questions, much of this material would not have been brought through.

I also wish to thank my editor and publisher Barbara DeBolt for her questioning mind and intelligent editing. She has kept me "on the ball" as far as outlines and editing and all the many details involved in publishing.

There are many others who have helped me in other ways, some by just being there when I needed them. I want to thank them all.

The cover design is also by my son, Neil. It is from Neil's original photograph of the moon and Venus taken from our backyard in Grand Junction, Colorado on Christmas night, 1995.

CHANGE

Lyrics from song "Change" by Norma

Rain is falling, trees are swaying
Leaves are glist'ning in the dawn
Sky is gray now, air much colder
Branches blowing oe'r the lawn.

When it's time the season will change
And our lives won't be the same
Where it's warm now, will be colder
And there's none but us to blame

Season's changing, life goes onward
Bringing much to think about
Walking slowly, down the pathway
Wond'ring how to end the doubt.

But I know it has to change
Yes my life is moving fast
Ev'ry moment leaves me reeling
Swirling mem'ries cannot last.

Where's it going, life that's moving
Blowing reckless here and there
Always onward, ever forward
Circling upward like a stair.

Yes I knew that it would change
Everything is changing fast
Every moment now is precious
Fading mem'ries from the past.

Caring, sharing, learning, growing
Brings it all straight down to me
No more leaning, never clinging
Standing on my own you see.

Yes the world just had to change
And the change came very fast
Every moment, things are different
A better world is here at last.

No more gray skies, sun is shining
Bringing music in place of tears
Laughing, loving, gladly giving
We are in the golden years.

Yes we knew that it would change
That the old would pass away
Though it had much pain and sorrow
A shiny future's here to stay.

Introduction

There is much confusion today concerning predictions of imminent cataclysms. Many people have been picking up information about cataclysms from within themselves during meditation. For some this information has been as memories about the long journey planet Earth has been on. For others it has been as visions into the future. Because of this it is very difficult to tell which is which, when standing in the middle of them.

This is a part of the spiritual path that many people will go through, especially if they had previous lives in Atlantis. At one time, about fifteen years ago, I believed in these predictions of coming destruction so strongly that I published them in two books, "Truths of Man's Divine Heritage," Volumes I and II.

Part of the problem is with the fact that when a person sees a vision of an earthquake or experiences one in real life, to them it is a devastating cataclysm, while in reality it affects only a small part of the planet at one time. The devastating cataclysms that many are contacting in meditation are memories of earlier cataclysms which were world-wide cataclysms and affected the whole planet.

Therefore, even when a psychic today has a vision of a "devastating cataclysm" there will not be world-wide destruction all taking place at the same time as in these other events. Even if California experiences the "big one," it will still not be in the category of "wide spread destruction" that many people are predicting. In reality it will affect only a small percentage of the overall population and land surface of the earth.

A new planet has been attracted into orbit around our sun and all the planets in our solar system will have to adjust their positioning to allow room for this newcomer. This new positioning of our planet will be a gradual realignment, but will cause many changes on the planet. We will still have earthquakes and destructive storms as part of this process, but these should not be forecast as destruction of the earth. Once Earth is settled into its new area all will return to normal.

Once the move is made, the new energy coming from this slightly higher, virgin area of space will allow humanity to reach higher levels of consciousness. The immediate goal is the transition from third dimensional consciousness to fourth. This will be a releasing of old emotional reactions and replacing them with compassion, love and joy. In this book I hope to explain not only what is happening to our planet, but also why so many predictions of destruction have been made in the past and are still being made. At the same time, perhaps this book will alleviate the apprehensions of many by showing the goal of the fourth dimension and also help us realize that we cannot reach this goal until the transition of the planet itself is finished.

Norma

Illustrations

Table of Contents

Part I
Cataclysms Past & Present

Part II
Cataclysmic Behavior Cycles

Part III
Personal Cataclysms

Part IV
After the Cataclysms

Part I

Cataclysms Past & Present

Chapter One:
Earth's New Orbit

There is much concern in today's world about the increase in destructive Earth changes and weather patterns. Many people are apprehensive because we are getting heavier than usual rain in certain sections, causing floods, heavier snowfall for longer periods of time in others, and drought in other areas which is a major cause of wildfires. The hurricane seasons are longer and are spawning many storms that are even more destructive than in the past. Everything seems to be going awry. The heat waves in summer are much hotter and last longer; the cold spells in winter are longer and colder. There are more tornadoes in not only expected areas, but also in new, unexpected places.

These changes are a sign of Earth's life forces awakening in preparation for the move into a new orbit that it is making. The planet goes through cycles of change, and at the present time we are in a more intense period of activity of this nature. As stated above, our weather is becoming erratic, our seasons are changing, and climate conditions are turning around. The plates beneath the planet are moving, bringing about earthquakes and causing inactive volcanoes to stir into life and then erupt.

This move has been taking place for so many millions of years that it is not detectable to our astronomers. It is a similar situation whereas those of us living on the land surfaces don't feel or realize that we are spinning and moving through space. Another example is that our continents are in a constant process of movement but we would not know this or realize it if scientists did not tell us these facts. Earth is moving through space in a slightly different area each time it revolves around the sun. In fact, it is constantly in a state of changing its orbit. Nothing is ever stationary in the aftermath of a nuclear explosion. As with the movement of our continents, we cannot know this without being told by a scientist. This is what the beings in the higher planes are, scientists, and they are telling us facts that we cannot have any way of determining ourselves on our planet.

As given, our planet is moving into a new cycle of space due to the birth of a new planet in our solar system. I'm referring to it as a cycle of space because the large universe is based on cycles of rhythm, which are cycles of space rather than cycles of time as we relate to cycles. Everything is in a state of change due to the planet entering this area of space that is resisting the foreign object being thrust into it by forces of rhythm in space. The actual move itself is creating more vibrations, which are causing the internal plates to move, bringing about earthquakes. The sloshing around of the molten lava is causing volcanoes to stir into life and then erupt.

To help you get a picture of planet Earth going through this move, I would like to compare it to moving a house from one location to another.

Several years ago my husband's brother and his family in Ohio decided to build a new house right next to their old farm house. When the new house was finished, the old house was loaded onto a truck and trailer and moved down the road several miles to a new location. If you have ever watched a house being loaded onto a truck and trailer and moved down the highway to a new location, you realize that many things can go wrong once this move begins. The most dangerous times are when the house is being put on the trailer and when it is being removed from the trailer. But many things can happen while it is on the road. Picture, if you will, what would happen if a tire were to blow out, either on the truck or the trailer.

This happened recently on a Phoenix freeway. A manufactured home was being moved on a truck and trailer, and the trailer had a flat tire. The rig pulled off to the side of the freeway and while the man was changing the tire, an eighteen wheeler went by. The suction created by the force of the wind from this passing truck caused the house to roll over, killing the man changing the tire.

This accident was not brought about by anything having to do with the house, other than the fact that it was being moved, but it certainly affected not only what happened to the house but also the man moving it.

In comparing this move to the one Earth is making picture, also, a weak wall in the house that begins to vibrate when it is pulled over rough ground. It could possibly collapse. Even a window could fall out if the vibrations became too great, which would happen if the speed of the truck moving the house were too fast. Other things could happen. The house could come into contact with electric lines across the road that someone forgot to move. This could cause unforeseen problems which would delay the trip down the road to the new location. The ties holding the house on the trailer could become loose and perhaps another piece of equipment would have to be brought in to help hold the house steady until it reached its new location.

Can you imagine what would be going on in the minds of the family that is following closely behind the house and worrying about each detail of the move? The children, while being fearful the house will fall apart are, at the same time, in a mood of being on vacation.

They feel they can get away with actions they couldn't get away with before because the minds of their elders are on the problems associated with moving the house. The same things are happening to the planet and her inhabitants.

If you've ever moved, you know it can be an exciting yet rather scary time. You're apprehensive about the new home and neighborhood and, at the same time, sad at leaving the old familiar surroundings. If just moving from one house to another is traumatic, think what a big job it would be to move the entire house!

My brother and sister-in-law made a video of the process of moving their house beginning with emptying out the old house, discarding excess baggage, securing everything to the best of their ability and loading it on the trailer. They then moved it very, very slowly with people walking along each side and in front of it while it moved and others leading and following behind in cars. When it reached the new location it was taken off the trailer and settled onto the foundation that had been prepared for it. Much rebuilding and repair had to be done at the new site, including running in electric lines, digging a well for water, grading the land, putting in a driveway and landscaping. The house needed to be painted, some windows needed to be replaced and the chimney needed repairing. All of this made a fascinating video to watch.

Let's look at this video they made while they were preparing to move from the location they had lived in for a long period of time, close to forty years. They go through a process of discarding things they do not want to move to the new location with them. They put things in boxes, reorganizing them as they pack them and marking the boxes. These preparations go on for a lengthy period of time before the physical location changes.

Besides all of the physical activity, they are mentally preparing themselves for the move. They will miss the preparing themselves for the move. They will miss the old place. They have lived in it for a long time. Many memories are associated with it, some good, some not so good, but nevertheless they will miss the old homestead. At the same time they are anticipating the move to the new location. Will they be comfortable there? Will they like the neighborhood? Will they like their neighbors? Will the garden grow as well? Will there be a lot of changes to get used to? They know the answers to all of this, but still they go over and over the move in their mind

Even the children in the house are upset, not so much because they don't want to leave – they haven't been there that long and don't have as many memories to contend with. Their upset is coming because they are picking it up from their elders – their parents and siblings and also the neighbors who don't want to see them leave. Therefore, the behavior of these children becomes almost unbearable. They are afraid because they sense the fear of the unknown in those around them. They demand more attention and commit acts to draw this attention to themselves.

The same thing is taking place with the youth on the planet today. They sense an uneasiness in their elders, a feeling of unsettlement and upheaval because their lives are going to be changed when the move to the new location is made.

Most of you can relate to what goes on as the move approaches. Now picture all of the same actions and reactions taking place on a much larger scale as the planet prepares to make a move to a new area in space. In many ways it can be compared to my brother-in-law's family. The same things take place. Things are discarded and cleaned up, changes are made, careful preparations are undertaken.

The emotional state of the people who are preparing to move from the old location becomes more unpredictable as the time draws near to leave it. Some are dreading the move, some are anticipating it, but all are in a state of emotional turmoil.

The same is true of the inhabitants of a planet that is getting ready to move. They make promises not to neglect the planet once it is in its new location the way they neglected the old one. They promise they will keep the area around the planet clean and neat and keep the planet itself much cleaner than they did in the old location. It's like having a second chance and they are going to make the most of it. Their intentions are very high, before the move is made, to do everything exactly right in the new location. They will take care of things that need to be fixed before they get to the point of not being able to be repaired. They will even watch their own attitudes and their reactions to others they are living with on the planet to make sure it is a peaceful place to live.

In my brother-in-law's case, moving the house to a new location was very necessary due to the new house being built within a few feet of the old one. To use another analogy say, for instance, that the area a family is living in now has become dirty and polluted and crowded. The fact of being crowded is one of the most important ones. Say, for instance, a new housing development has been built within a short distance of the old house. There is no longer room in the neighborhood for all the children in the school or the cars on the street. In both of these examples, my brother-in-law's house and our "pretend" crowded neighborhood, they are literally being pushed out to make room for the new development.

The same thing is taking place in the large universe.

There is a new development in the neighborhood of planet Earth and she is being pushed into a new location to make room for this new planet. The physical changes are cleansing preparations before she makes the move to her new home. She has already started to make the move.

This move has actually been taking place since the time of Atlantis when there were very destructive cataclysms. In our analogy, this was when the planet was lifted up onto the "trailer." It was a perilous time and almost disastrous. We had much help from beings on the higher planets at that time. They stayed beside the planet and steadied it as it was being lifted onto the "trailer" and have been beside the planet slowing her move and easing her path in any way they could. They are still with us now as the time approaches for the planet to be lifted off the "trailer" and settled into her new location.

Picture it as though the inhabitants had refused to leave the house and were still inside when it was being lifted. It tilted over onto one side and caused much internal destruction to the house and to the inhabitants. The inhabitants were taken off and not allowed back into the house until it was secure, and then only a few special ones were allowed to ride with the house on its entire journey. This would relate to the number of people who perished at the time of the cataclysms in Atlantis. There has been an ever increasing number of inhabitants allowed to reenter the house while it has been moving into the new position.

The planet is almost at the new position and many who were taken off when it was first lifted onto the "trailer" are now back on board and will be with it as it is lifted off the "trailer." They were trained in how to help with lifting the house from the "trailer" and getting it settled into its new location. They have many ideas about new ways to heat the old house, grow food for the family, and to train people as to how to keep the new neighborhood from becoming as corrupt and dirty and polluted as the old one was.

To compare the stage our planet is in at the present time to our house analogy, we are in the process of being lifted off the "trailer" and positioned in the exact new spot. In the large universe this means the planet has to rise just slightly higher to reach its new orbit. Once our planet is secure in the new location the tempests will let up, but first it will need to make a few revolutions around the sun to ease smoothly into the new path.

When the planet is finally at the new location, the entities from the higher planets will serve in the same manner as the men and their equipment who helped steady my brother-in-law's house when it was being taken off the trailer. In the case of the planet, this will all be done by cosmic waves of energy. These waves of energy can be manipulated by the entities on planets in other solar systems and from the capitol world of our solar system. This manipulation and controlling of the waves of energy that are propelling the planet into its new position, is done with thought or electrical energy. This thought energy is being used as a dam to slow and control the flood of energy (made up of subatomic particles) that is moving in a tidal wave towards our planet. These entities, our space brothers and sisters, are doing their best to help the planet make a smooth move. They are also increasing the vibrations of the planet very slowly and holding it steady by magnetic pulls at strategic points on the surface of the planet.

This may sound like science fiction but indeed it is not. There is a whole world of activity in other dimensions that we cannot even begin to understand that is helping Earth make this move to the fourth dimension. These entities on other planets in other solar systems went through moves like this and know what to expect. In other words, they are experienced "house movers."

When our planet has made the move, these same entities will help the next planet until the whole solar system has settled down and made room for the new planet. In our analogy of moving the house, this would be similar to the old house reaching its new location and settling into its spot and the family safely living in it again.

Both the planet and its habitants must make the rise to

the fourth dimension at the same time. It can be no other way. Earth cannot rise up in vibratory rate into her new orbit unless those living on her do also. On the other hand, we cannot rise up in vibratory rate without the help of the planet. Evolutionary growth is always a reciprocal action - always.

The solar system will eventually have twelve planets but it will only have nine planets at any one given time. There will always be one that has completely moved out of orbit and two ready to move in. Actually there are two moving out. Let me try to explain. When the first one dropped out, the tenth one was added for the total of nine. At the same time that planet number two is in the process of dropping out, planet number eleven is in the process of moving in, with planet number twelve waiting in the wings.

This is the exact stage where the solar system is at the present time. Planets eleven and two are operating at a bonded rate of frequency, as did planets one and ten when planet number one left and planet number ten, Pluto, entered. The problem is coming because of the lagging behind in time of planet four - Earth. It is not able to make the move into the new orbit as quickly as it should to allow planets two and eleven to stay on schedule. Planet Earth really is holding up the progress of not only our solar system, but the whole galaxy and the large universe. The only way the planet can rise in vibratory rate is by its inhabitants raising their consciousness from third dimension to the fourth.

Each person who does not raise his consciousness to the fourth dimension is actually hurting the progress of the Creator. It is awesome to think that every single one of us has the power to hurt the progress of the large universe, but we do.

There is a certain frequency to the orbit of each planet – a preset frequency. Each one moves one step closer to the sun when a new planet enters. It is then consumed by the sun and enters a black hole to be rejuvenated and reborn at a later time in another solar system, one of the new solar systems of the twelve planned for in our galaxy. No energy is ever totally lost. Our planet is out of tune with those ahead of it, thereby holding up the tuning of those behind it.

I'd like to compare the process of tuning our planet to the process of tuning one of our musical instruments. We don't have a nine-string guitar but let's pretend for a minute that there is such an instrument. A guitar is tuned starting with the first string, the finest string. This pitch is set by absolute pitch, say from one of the new electronic tuners on the market, in the case of the guitar. In the case of the planet it is set by a master tuner, the sun. The pitch of each planet is determined by the atmosphere around it, which is made up of energy waves, or the solar wind, coming from the sun.

All the strings on the guitar are then tuned to be in perfect pitch according to the pitch of the first one, whose pitch is pure and accurate. If there is something wrong with one of the strings that won't allow it to be brought up to pitch – say the tuning peg is stripped or the string is old and has lost its tension or some other reason – none of the rest of the strings can be tuned accurately until this one string is fixed and pulled up into the correct pitch. The correct pitch is determined for it by a mathematical formula which will then assure harmony in the chords to be produced by the whole guitar. In the case of the planets it will be the harmony of the spheres to be produced by the whole solar system.

Part of the problem with bringing this string into pitch is corrosion of the guitar string. In the case of the planet it is the dirty atmosphere around our planet. This is the exoteric or physical reason for the problem with either the guitar or the planet. The other part of the problem is a esoteric or mental reason. In the case of the string it is the inability of the person doing the tuning to hear the correct pitch. In the case of the planet, this would relate to those who refuse to open their minds to hear the new messages coming to them from the Creator.

The first string breaks, as first strings are inclined to do, when it is pulled too high. In the case of the planet, it gets too near the sun. Instead of putting a new first string on, because we do not have a new first string, all the other strings are moved over and the tension on them raised and a new ninth string is put on because we have one of those in our repair kit. As they are moved over, or into a higher position on the guitar, all the strings have to be pulled up in tension to attain the pitch of the string that had preceded it.

Anyone who knows about guitars knows that this is quite difficult to do. The strings weren't manufactured to be able to do this. But let's pretend for a moment that there has been a new substance brought into being to manufacture guitar strings out of – a new alloy. This new substance allows for this tightening of tension to take place. In the case of humans, this would relate to the new atomic structure of human forms. The guitar string will still need worked with and looked after, as far as tuning, for a considerable period of time before it will hold the new pitch, but it eventually will. The same with planets. The new strings are made to do this and to also, then, be able to be moved to the next highest string and be tuned up when the time comes. The same with planets.

This is the beauty of this new nine-string instrument. The strings can be constantly raised higher. Then they finally wear out and break after having served their usefulness in first position. Because the instrument itself is so beautiful in its harmony, when each one is ready to move up it attracts just the right new string into its "locked in" harmony. This will be at the lower vibratory side of the guitar, the heavier string. You see, this guitar is almost like a living entity in that it can and does control its own progress.

As far as planet Earth goes, there has been a hitch in the tuning of this planet as it is being moved to the next higher position. It must make this move. The planets ahead of it are tired, the first string now – that previously was the second string – is very tired and weak and ready to be relieved of its duty, but it cannot be removed until the problems with the string that is planet Earth are resolved. The only way these problems can be resolved are by the inhabitants of this planet.

I hope this analogy to the guitar helps you understand the position we are in at the time. The planet is trying to do what it is intended to do, but it cannot be successful until its inhabitants do their share.

Chapter Two: Cataclysmic Memories

If you've ever moved you know there's a period of utter confusion after the move. Everything is in boxes. You can't find the pots and pans to cook in or the dishes to eat off and usually end up eating fast food. The bedding is packed away in the bottom of a box and you're tired, so you sleep on a bare mattress that first night. The next day you can't find clean clothes or the other pair of shoes you want to wear. You are still in a state of utter confusion. You begin to have dreams that are influenced by what you know must be accomplished. You might even regret that you ever wanted to move. Sometimes you begin having nightmares about the boxes coming after you, overtaking you and squashing you as a "bug" in their way. It can take weeks to come out of all this confusion that is influencing your dreams.

I went through a period of confusion concerning cataclysmic activity about thirteen years ago. At that time I started having vivid dreams about destructive cataclysms. I was taking a class in metaphysics at a local bookstore where they also held past life regressions for those involved in the classes. We usually had this done about three times a year. My first three experiences were normal regressions that showed a pattern of behavior following through each one. This was basically the goal of having these regressions done. The next time I had one done, the regression went forward instead of backward. I was progressed into the future where I saw a vision of Hawaii disappearing and the rescue of people on the mainland by spaceships. After the progression, those helping me tried to have me establish a time for this event, but there was no time line evident in the vision at all. Later I would understand why when I began to see the "light at the end of the tunnel." But at that time I went through this period not knowing what to think or believe.

The vision I received during the progression was a picture of what I, and many others, saw happening at that time (this would have been the early 1980's). This is the same picture that many are still seeing. You see, this picture was brought onto the Earth plane as a physical reality because at that time there was a possibility it would happen this way. Due to the fact that "thoughts are things," this picture was imprinted on a "compact disk" of memories in the conglomerate mind, by those on Earth who were contacting these thoughts.

At this point I'd like to give a description of what I saw

in my "vision." Please keep in mind that this picture is no longer accurate. Due to unexpected cosmic events it is not going to happen this way.

In my progression I saw that the volcanic activity that had started that year (1983) never let up. It continued until eventually all of Hawaii went into the sea. The survivors left by ship and headed for California. As they approached, they could see that the coast line of California was all chewed up, as though a giant had taken big bites out of it. I experienced the vision as a young girl named Manoa. The characters in the vision were merely used to enable me to relate to what was being seen, to make it more personally meaningful. The physical events are now imprinted on the "CD of memories," not the characters.

The sky was glowing with the light from the volcanoes spewing out flaming lava and rocks. Manoa turned and looked back at the old man following behind her. Her grandfather was very sad to be leaving the islands. His wife of forty years had perished in the initial eruption of Mauna Loa early in 1984. Kilauea was also erupting then. Now every peak was blowing its top. There was no way they could stay any longer. As Manoa walked along the beach, kicking the sand with her bare toes, she sensed more than saw others following her and her grandfather. They were all heading towards the ship that would take them off the island. All were carrying what they could in packs on their backs. They were all quiet; each had his own thoughts. None of them had ever thought it would come to this, that they would have to leave their beloved islands. The volcanic activity had actually never let up after it started back in 1983 and the island had become uninhabitable. Two of the Hawaiian islands had already disappeared beneath the turbulent waters of the Pacific.

They were getting close to the ship now. It was quite a large ship. Manoa had never been on anything larger than the outrigger canoes they used for travel between the islands. She brushed her long, dark hair back from her face and turned to wait for her grandfather to come up alongside of her. In the glow from the volcanoes, Manoa could see the tears slowly creeping down his lined cheeks. She linked her arm in his and gave him a squeeze of encouragement. "It's not much further now, grandpa," she said. "Let me carry the pack for a while." "No, I'm okay Mani," her grandfather said. "I'll make it. Don't worry about me."

Louan and his wife had taken Manoa to raise when her parents, their son and daughter-in-law, had died in a fishing accident when she was just a year old. She was now fifteen and had been a real blessing when his wife died. He couldn't have gone on if it hadn't been for her. Now she was frightened and upset at having to leave the island. He knew he had to be strong for her sake and he straightened his shoulders and wiped the tears from his cheeks. He hoped she hadn't seen him crying. "Everything's going to work out just fine, you'll see," he said to her.

The crew was busy loading the ship with provisions that would be needed for the trip. The refugees boarding the ship were carrying their belongings aboard and trying to stay out of the way of the crew.

After Louan had found a spot to set down their pack, their most treasured items because they couldn't take much, he moved over to the ship's railing. He stood there watching as a new eruption started. This one was much closer to the ship and he knew the time was short.

It would be less than an hour perhaps, until they would have to cast off or they would not be able to get away. He shivered as the realization hit him that this eruption had by now covered the house that he and Manoa had just left several hours ago. It really was a close call and he was thankful that Duon, their neighbor, a young man in his mid-twenties, had knocked on their door and insisted that they get out. If it had been just himself he probably would not have left, but he knew he couldn't do that to Manoa. He had to try and save her. They had quickly thrown the things they needed to have and a few treasures like his wife's picture into packs and set out. He saw Manoa approaching him and realized that she was barefoot. "Mani, where are your shoes?" he asked. "It's okay Grandpa, they're in my pack. I didn't take time to put them on," she said.

The sky was getting brighter now and the combination of smoke and steam was making it difficult to see. The crew was hurrying faster to make the ship ready to sail. Manoa and her grandfather stood clinging to each other and watched as a finger of lava further up the beach reached the water and sent up a large cloud of steam. That finger of lava cut that part of the island off and Louan hoped all the people were at the ship by now. He realized that the last ones to come aboard had to wade through water to get to the ship. This told him that the island was sinking and soon would be under water. Well, he thought, that was one way to put the volcanoes out.

The sea was choppy. Manoa didn't feel well. In fact, she had never felt so awful in all her fifteen years. She was sitting in a corner of the deck, very pale and shaky. She saw her grandfather approaching carrying a plate of food. Oh no, she thought and turned to the bucket that hadn't left her side for days. Louan put down the tray of food and kneeling next to her held her shoulders till the spasm was over. "It won't be much longer, little one. We should see the shore of California before nightfall." He sat down beside her and she leaned against his shoulder and fell into a light sleep.

The captain was confused. They had finally spotted land, but could find no familiar landmarks. By rights it should be California, but the land was much further east than it should have been. He had anchored the ship until daybreak and would try to get their bearings in the morning.

The captain didn't sleep much and at the first crack of dawn was up looking at the shoreline through the binoculars. The coast line was all chewed up, like some giant had taken large bites out of it all up and down the coast. The cliffs were very, very high. There was no way he could put the ship into land here.

"What the hell has happened?" he said to his mate. "Has the whole world gone berserk? I know the weather has been freaky and the volcanoes were certainly erupting strongly in Hawaii, but I never expected to find this kind of damage in the states. Call the crew together for me," he told the mate.

When the crew was gathered around him he said, "We have to change our course, men. Look at that coastline. We'll have to go down around the tip and try to land in Mexico. I don't want to alarm the refugees. They've been through too much already. I just hope we can land in Mexico."

The ship continued down around the tip of Baja, California and landed at Guadalajara, Mexico. From here the survivors went to Texas, but didn't stay long as Texas began to go under water.

It was as though the whole lower part of the United

States had tipped down into the sea. Panic was everywhere. The government was helpless; the churches were useless. The atmosphere was filled with a dense smoke that seemed a reddish color. There were about twelve areas around the country with groups of survivors with each group having from three hundred to five hundred people. All the groups seemed to be on top of mountains or up on high ground waiting to be rescued.

Please remember that this picture has since been changed. Between this vision of Hawaii disappearing and the vivid dreams I was having at this time of cataclysmic destruction, I was very confused. During meditation one day I asked for an explanation from the higher planes about the dreams. I was told that the dreams were memories of the cataclysms during the time of Atlantis and that any cataclysms coming now are still part of the polar shift that took place at that time (when the house was lifted onto the trailer to be moved in our analogy). That move is ongoing, therefore, the cataclysms are still ongoing, but not of as violent a nature (because it is up on the trailer moving down the road). Actually, the original nuclear explosion, or "big bang," is ongoing and we are still in the chain reaction of it.

The dreams and memories of Atlantis were from the past. They were third dimensional pictures which means that they had already manifested in physical matter. The vision of Hawaii was in the future. It was a fourth dimensional picture which means it was in an unmanifested realm in ethereal matter at the time I saw it.

Warnings of impending destruction were being received by many people who were in touch with the fourth dimension. There was a very real concern by the entities in the fourth dimension that the realignment of the planet into its new orbit was going to be very destructive. The fact that it was just a possibility instead of a reality was the reason why I couldn't put a time on the vision I had when the progression took place.

But fourth dimensional visions can be changed. The fourth dimension is the research and development phase of planet Earth. All things in a research and development stage can and should be changed. This is what R & D is all about. Nothing stays the same. Once the possibility of great destruction was seen, those beings in the fourth dimension went to work to find ways to prevent the destruction, or at least keep it to a minimum.

Because thoughts are things, this picture had already dropped to the third dimensional Earth plane and, being a very vivid picture, it was imprinted on the "CD of Memories" in the Creator's mind and became accessible for anyone tapping into their inner beings on a third dimensional level.

I'd like to attempt to explain the "CD of Memories" by comparing it to our method of making records or compact disks, CDs, here on Earth.

During the time I lived in Los Angeles I went into several different recording studios to make demo tapes of my musical compositions. I know most of these recording studios have the capabilities of recording at least eight different tracks or instruments while others have more. This means that you lay down a "track" for, say, the drums first. Perhaps this is followed by a bass guitar, then a piano, trumpet or violin and so on, laying the sound of each instrument on top of the previous ones. The last thing to be added, in most cases, is the vocals. Each recording is a composite of all the different tracks.

In comparing what happens in these recording studios to what happens in our galaxy, picture the ethers as very sensitive, impressionable material. This ethereal CD is more sensitive in outer space and less sensitive as you draw in towards the planets. For instance, in outer space a single grain of sand, as we know it on Earth, could tear a hole in this ethereal matter to the point of destroying an area comparable with a square mile of Earth. Of course, there is no sand as such in the great universe. The equivalent to sand or any other destructive item in outer space would be thought.

Because thoughts are electrical bits of energy, they

work with the precision of a knife to slice an area of ethereal matter. They are the ultimate "space weapon," if you will.

Thoughts, even mild, pleasant thoughts, can make an impression of the type that is needed for a normal CD in our analogy. Therefore, any thoughts stronger than a normal, everyday thought - any thought that, say, stirs up the emotions in a human - will cut a deeper impression in our CD. Any thoughts of emotion will do this, but an angry thought is far more destructive than a thought of sorrow, although a thought of sorrow will groove deeper than an ordinary thought. A thought of fear is the most devastating of all thoughts, as far as grooving an impression on our CD. This is also true as to the effect it has on a human being. (Think of the fear felt by all those who perished at the time of Atlantis and the impression it made on both the CD and the humans.)

Picture this CD we are talking about as the Milky Way Galaxy. This is the holding place of the mind of the Creator God over our solar system and other solar systems in our galaxy. The surface of this CD is marked off into twelve blocks, or sections, one for each of the twelve solar systems in this galaxy. Each block or section is then divided into twelve tracks, one for each of the planets in each solar system. When the Creator moves from one solar system to the next, His mind is a composite of the twelve tracks made by the planets that made up the previous solar system. As you can see it is a very large CD. The Creator will be a very wise Creator when He has finished His trip through our galaxy.

Let's narrow this down and focus on the track for planet Earth. It is the fourth track in the block set aside for our solar system. The whole CD is all the Milky Way Galaxy but this is the one spot that shimmers for us when compared to the rest of the record. This is the holding block of the Creator's thoughts for our planet. Because the Creator's mind is the same mind that is in us, it is a block of thoughts coming from humanity itself. It is called the conglomerate mind. The thoughts of humanity make up this mind only as far as things pertaining to our planet.

To explain, the inner being that all humans have access to is a conglomerate of all thought that has ever taken place at any time throughout the history of the planet. This conglomerate inner being is not the divine mind. It is human mind condensed and compressed into only that which serves to further the growth and evolvement of human beings. There is an elaborate process that takes place with the thought-forms that stem forth from humanity. For the most part these thoughts are contained close in by the planet until they have been proven to be of absolute value as far as furthering the spiritual uplifting and uprising of the human spirit out of the animal flesh.

The thoughts that go forth are monitored and filtered into different conglomerates. These conglomerates are made up of previous personalities of those of us who are incarnated on Earth at this time. They take the current thoughts that are coming from us and work with them and send them back to those of us who tap into this conglomerate mind. Then we take the thoughts, send them out through several different means: books, lectures, etc., to other people who, in turn, have their thoughts to add to or change what they read or hear. These thoughts are then automatically sent back to the conglomerate mind where they are once again monitored and filtered.

These thought concepts that go through many periods of back and forth growth are what make up cycles in human history. They are also what is used to determine lessons that are to be worked on when a person begins a new incarnation. This conglomerate or inner being of all humanity from time immemorial is the one, single most important factor or way of producing spiritual growth - and thereby fostering new stages in evolution - that is available to the planet. It has not been understood up until this time, not even remotely by any. Our schooling should bring forward the inner being of each of us at a very early age. This will lead to the necessary intertwining of this inner being with our present personality. This is most important. The direction that some people are taking now is that of turning our whole incarnation over to our inner being.

This is not what is wanted or needed. Can you imagine what the world would be like if all these inner beings were walking the face of the planet in embodiment? There would be no progress made because we would be stuck in outdated ideas for the most part. The way our inner being can help us is for us to be on the Earth plane experimenting and sending them the results of our experiments. Our inner beings then take all this material and condense it and come up with overview thoughts. These thoughts then are sent back to us for experimentation and this, then, returns to them. It is a cyclic ebb and flow of information that will progress the human race. If our inner beings were walking on the planet with us or walking on the planet by themselves we would not progress. It takes those of us down here on Earth returning to those up there.

It was tried the other way and did not work. This was what Atlantis was all about. Since the restructuring of the human form that took place after Atlantis, progress has been much faster. Communication with these inner beings has started opening in so many that progress is much, much faster and becoming more so all the time. Eventually this communication with our inner beings will be wide open on children as they are born. This does not mean that they will remember their past lives, only that direct communication can be had with their inner being in a momentary, cyclic process. This is how your ability to communicate with your inner being should be trained to work. This is the proper way. I wanted to explain it so you would know the proper goal for all. You are to be able to flip flop from you down here to those up there at any given time with no noticeable effort on your part. It is a mental process. They are in your mind. Think of it as a tiny door and see the door open when you want to access them.

This is the process that makes the grooves for planet Earth on our CD. The rest of the space on the record is the remainder of His mind, for use by the other planets in our solar system and for other solar systems, some of which are not formed yet.

The fourth track out, where you will find the shimmering spot set aside for us to imprint thoughts onto, is fairly near the hole in the center of the record. The further out towards the edge of the record you go, the space belongs to the newer, unborn solar systems.

To review, the block allotted to our solar system is divided into twelve tracks for twelve different songs, one from each of the twelve planets in our solar system. We are on the fourth track out in the shimmering block allotted to our solar system. The first three tracks are finished. These were songs of first, second and third level consciousness, respectively. We are in the recording studio attempting to lay the tracks for the fourth song on our block. This will be of fourth dimensional consciousness level which is the ability to be in contact with our inner beings.

Some artists, when they go into the recording studio, have an idea of what songs they want to put on their finished record. We also have an idea of what kind of song we want to put on our block of the record as the fourth song, but are not quite sure as to how to do this. We are still thinking thoughts similar to the song on the third track. We need to break out of the influence of that third song and focus on a new song for the fourth track. We are not at all sure what songs will be on the tracks after the fourth one, either. Sometimes it works out as planned, but more often than not it is all changed by the time the recording is complete.

When humans tap into this shimmering spot in the conglomerate mind, they begin their contact with the one particular track allotted to our planet. This contact is then progressive. This track was started when the Planet was formed and every event since then has been impressed on it. The more emotion attached to a thought, the deeper the impression and the more powerful that emotion is, the deeper the impression will be. The thoughts of each individual are on this track but are infinitesimal in size.

From this point on I would like you to visualize this one particular track allotted to planet Earth as an upright pole. See the track being lifted off of the compact disk and stood upright as a pole, with the forming of planet Earth at the bottom of the pole.

As humanity was born and then died, these memories were stored as pictures on the CD or "pole," if you will, in the brain stem. The stronger, more frightening and cataclysmic the event, the stronger the picture or memory of it is in the conglomerate mind. Also, the more humans who experienced an event such as this at the same time, the stronger and larger was this "record" on the pole of information in the conglomerate mind.

Many, many thousands of people died at the time of not only Atlantis, but also during other cataclysmic events throughout history. Some of these events were stronger than others and some of them left stronger memories simply because of the evolutionary growth of the people that were leaving the memories. Therefore, even though the cataclysms at the time of Atlantis weren't as destructive as the ones that took place when Pluto entered our solar system causing the movement of glaciers, etc., the Atlantean cataclysms left a much stronger imprint on the pole in the Creator's brain stem. This is the picture of cataclysms that is usually accessed by most people at the start of their inner journey because it is larger and stronger and more recent. Besides people tapping into these events that are stored as group memories, they also can tap into individual memories. In fact, some people, if they died a particularly horrible death from a cataclysm will most likely tap into the individual memories first before the group memories.

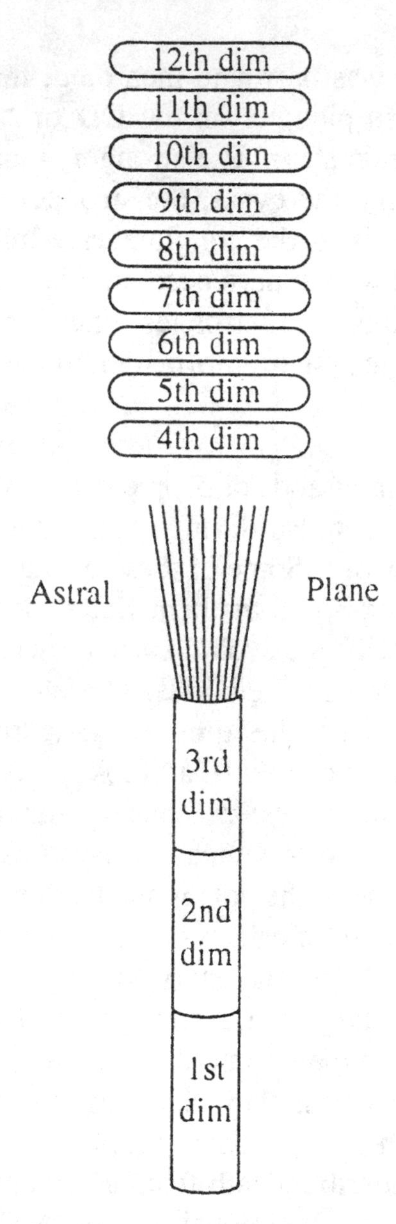

The "Memory Pole" in the conglomerate mind showing the astral plane — Illustration by Neil Hickox

People who were leaders, such as Jesus, left very strong imprints on this pole, individual imprints involving many people. The whole Biblical period left strong pictures on this pole. This is why so many people access Jesus and Mother Mary when they begin connecting with the conglomerate mind.

As people finished their cycle of reincarnation on Earth they progressed onward and upward to the higher planets, filling up the track allotted to each planet on the large CD. When picturing the whole block for our solar system as an upright pole, as in the drawing on the preceding page, you can see the area between Earth, which is manifested in physical matter, and the higher planets which are in ethereal matter. This area is called the astral plane.

Until you move past the "pole of energy" for planet Earth and move through the astral plane, the dangerous stage between the pole and the "rings of energy," you will not be accessing the higher realms. These higher planets consist of individual memories of "real entities" instead of group memories.

For those of you who died physically at the time of Atlantis, the records of cataclysms are the brightest spot on the pole of information. You relate to it because if you have not been back on Earth since then, you have not experienced the events that come after this on the pole of memories in a physical body. Therefore, when you begin to tap into the inner realms you hit this spot about cataclysms and it triggers fierce memories in you.

The period between accessing the pole of memories and the "live" communications from those in the rings of energy, the period of progressing through the astral plane, is the period when you are deeply immersed in psychic phenomena. This causes chaos in your mind to one degree or another, which then causes you to go through upheavals or cataclysms, if you will, in your personal life.

These cataclysmic events could be in the form of the loss of a job, a serious illness, a financial loss, a divorce or any number of other things. These events then have the tendency to set you firmly on the spiritual path. You will find that almost everyone can pinpoint a certain event in their life that caused them to become a spiritual seeker.

Eventually, I began to realize that everyone followed a pattern in the information they received about cataclysms. This was due to what had been imprinted on the pole of memories in the conglomerate mind that people connected with while in meditation. They were tapping into memories of previous cataclysms in their seedcore. Some of these memories were of the destruction at the time of Atlantis, because many souls returning today had lives in Atlantis and did not survive the cataclysms. These people were removed bodily (died) and learned new methods of doing things on other planets.

Incidentally, we no longer have to do it this way. We can be taken to the higher realms at night while we are on the inner planes and receive our training this way. Regardless, they have returned (reincarnated) now to implement these new methods. If this is the first time back on Earth for those who died at the time of Atlantis, those memories are still very vivid. A lot of the material about horrible destruction comes from those of you in this early stage of connecting with your inner being. These memories are very real to you at the time you are receiving them; therefore, you cannot help believing them. Once you move on and realize that no cataclysms like that have happened recently and that it doesn't look as though they are going to happen any time soon, you will realize that you were experiencing memories. Then you will start to look for answers about this and to other questions you have. This will cause you to grow and continue upward in your efforts to connect with your inner being. Everything that is on the pole of memories in the conglomerate mind will cause this growth to take place.

The same thing is true in an individual's brainstem. Memories from past events in not only the present life, but all previous lifetimes, are on the pole in the brainstem as far as that soul's experiences, and in the kundalini as far as the cells making up the physical form. It is all there to continue to foster the growth of the individual. Evolution demands that this growth take place. It must. The soul will see that it does, no matter if it has to "trick" the present personality by having it believe things like horribly destructive cataclysms being imminent.

Believing the picture of terrible cataclysms has always been a stage of spiritual growth. Some people continue to grow and move past it. Others get stuck here and devote their whole lifetime trying to find ways to survive. Still others convince large groups of people that the world is going to end tomorrow - doomsday people. These people should not be kept from their lessons, but at the same time they should not waste their lives trying to figure out the best place to be to survive cataclysms and the best preparations to make so they will survive. If the whole lifetime is taken up with this, what will become of the goals they set for themselves?

Much of the inner path is formatted by each individual to make spiritual growth take place. If you have laid out your path to get stuck on the "cataclysm step" of the stairway, there must be a reason. Therefore, you either will not read this book or simply won't believe what is in it if you do, but at least you will be offered a chance to grow into the reasoning intelligent mind.

As discussed, you actually have a double line of memories, the physical and the spiritual. In order to access and experience what is going on in a balanced, true fashion you must access both of these lines to come into the truth of a situation. Those only accessing the physical will bring through an unbalanced view and those only accessing the spiritual line will likewise bring in an unbalanced view. This is why synthesization of both parts of a human is so vital to having a well-balanced human being.

Reaching this stage of balance is very difficult for humans and most people don't persevere enough in one lifetime to reach it. This is dragging out the process of moving on for a much longer time period than it should take. This is throwing off the rhythm of the universe and will cause other events to collide because of this slowdown in the evolutionary process of those on the Earth plane. This classroom has been a sticking point for almost everyone and is becoming more so, and will become even more so, until everyone has an opportunity to understand their spiritual heritage.

To reiterate, the cataclysms were seen as being imminent by entities in the fourth dimension and therefore pictures of them were brought onto the Earth plane by people in contact with the fourth dimension. The possibility of these cataclysms taking place was then changed on the fourth dimension, to save the planet, before they actually manifested in the third dimension.

When I saw the pictures of Hawaii disappearing into the sea and the rescue of groups taking place on the mainland, they were fourth dimensional pictures or visions. At that point, the entities that were in the fourth dimension (former incarnations of mine and also of other people) thought that the shift of the planet was going to cause violent disruption. The picture of what was slated to happen had not yet been changed through the efforts of these entities enlisting help from entities in even higher planets. As this help came forward, the possibility for devastating cataclysms began to change and those of us contacting the higher dimensions began receiving a different picture, a picture that no longer included imminent cataclysms.

It is very important that we understand the difference between the two different interpretations of material concerning cataclysms. As given, there is a view from those still of third dimensional consciousness who are accessing the memories or pictures that were left of the original view of what was slated to happen. There is a different view from those already in fourth dimensional consciousness. The picture that is in the fourth dimension now is that of the cataclysms not being imminent.

The cataclysms simply are not going to happen as many people are still picturing them. The outlook is not so severe; the shift of the planet is not happening as fast as feared. It is a much slower process of moving into the new orbit. The whole procedure is taking place in a much calmer and more controlled way.

Chapter Three:
Predicting Upheavals

As discussed in the last chapter, there are two different ways that the memories of cataclysms are tapped into: through the kundalini or through the brainstem, a double "whammy," so to speak. Memories available to the physical body through the kundalini give a third dimensional perspective. Those obtained through the spirit from the brainstem give a fourth dimensional perspective. This makes it very difficult to tell where predictions of cataclysms are coming from. I would like to help you understand how we see things through two different perspectives by explaining how the process of tapping into these memories works.

Do you remember how long the summer seemed when you were a child? It felt as long then as perhaps a year seems to you now. I can remember returning to school one particular fall after having the summer off and everyone seemed different; everyone had grown up. I didn't recognize a lot of the kids and was shocked when they said who they were. I believe that, up until then,

I recognized everyone because I was relating to them strictly through physical remembrance.

That particular year it wasn't the other kids who had changed so much over the summer. I had gone through a change of some sort, and when I returned in the fall I'm sure those other children didn't look that much different. It was my perception of them that had changed. I was no longer seeing them just physically; I was accessing them through another part of me, an inner part that had somehow changed, rather drastically, the way I saw them.

This is similar to what happens when you begin tapping into your inner being. In the beginning, you are accessing memories that relate purely to physical evolution – things that took place in the past as far as humanity itself is concerned. You will relate to physical things, such as aspects of nature that are a part of the physical make up of every human being. Everyone has these memories as part of their divine heritage. But when you begin to tap into the realms beyond the third dimensional Earth you begin to access a different kind of memory. Suddenly your memories are different from anyone else's. They become very personalized memories that only you could have because you are now accessing an inner part of yourself coming from spirit that includes all your past lives.

No one else can have these memories. They are a combination of what you remember physically simply because your physical form evolved through all kingdoms and the lives that your soul spent in all the different physical forms. This is not merely accessing memories of your soul on the inner planes. These are memories of your soul reacting through different physical circumstances. Even though all these physical forms had the same beginning, there were many differences between them which influence the lives your soul will be able to live in this body. There was a difference in sex, race, nationality, intelligence and physical stock inheritance. But even with all these differences, certain traits and tendencies that are a part of your soul will still show up.

As an illustration of this, in the previous chapter I spoke about having past life regressions performed and that the first three regressions were normal regressions showing a pattern of behavior. For instance, I accessed one life as a lonely male, a herder of sheep. I don't know what country it was in but it had high mountains. In another life I was in Indian country in the early settlement days of the United States. I wasn't an Indian but a young girl, about fourteen, whose family was killed by Indians. I was rescued by a circuit rider preacher/judge, who took me to live with him as a housekeeper and to sleep with him. He then kept me more or less a secret from those he associated with and I ended up living another lonely life. Another life was as a female working in a theater as a janitor and watching those on stage and then going home to my room where I lived all alone.

As you can see, each physical body and life circumstances were entirely different, but there was a thread running through all of these lives, and even more of my past lives, of being a loner - keeping to myself and not mixing with others. This can readily be seen as a spiritual thread in all these lives that changed and colored my perception of how I saw people and events in those lives. The tendencies to behavior like this are what we incarnate to overcome. If it were not for the physical memories of the lives in the different bodies showing us that the same tendencies to behavior can manifest under any conditions, we would not be able to pick up this thread and take steps to correct it in future lives. This is basically what making the transition to the fourth dimension is all about.

To get back to explaining how this works in the overall scheme of things, not only can we experience past lives from a viewpoint of the third dimension, but we can then receive an understanding of the consequences of those same lives when we perceive them through the fourth dimension.

When connecting with the conglomerate mind, those people who are still tapping into the CD of memories for planet Earth - instead of the rings of energy, which are the higher planets - are receiving pictures of cataclysms that have happened in the past. They are accessing these memories through physical evolution. I would like to call these third dimensional pictures of cataclysms. This will distinguish them from pictures people see when they progress through the astral plane and access the fourth dimension. Here the memories are accessed though spiritual evolution and are the thoughts of those who have progressed past planet Earth. We will call these fourth dimensional pictures of cataclysms. These are seen by entities on the higher planets – who are, in reality, our inner beings, our former incarnations who have moved on – as possibilities for what might happen on planet Earth and are then passed down to those with whom they are communicating.

As given, everyone has a double set of accessible memories. Not only do we all share the same mind, that of the Creator, but our physical bodies also came from the same beginning process of evolution. All physical bodies evolved through the different stages of physical evolution. Some of the visions and pictures and forecasting of cataclysms are coming simply because of this fact. The memory of the very beginning of the planet, in its pliable, molten stage, is built into every human physical form, due to everyone coming from the same, identical beginning. Even with the many, many branches off of this beginning, the memories of the formation of the planet – which was still going on at the time that man was separated from the animal kingdom – are stored in the kundalini or root chakra of each of us.

These memories or pictures of volcanoes and earthquakes and floods and fires are available to each and every one. They are part of the brainstem from which the first human brain was fashioned. They are also accessible for the spirit when it enters a body.

To review all that we have talked about, let's pretend that we have been to a theater and saw a double feature movie. There was also a cartoon showing.

The first film of our double feature movie, "The Coming Cataclysms," was a very dramatic movie about the possibility of severe destruction on Earth. It must have made a very strong impression on those who saw it because the pictures of cataclysms reached many people in that time period. This was at the same time that I saw the vision of Hawaii.

I know it made a strong impression because I had friends that kept the walls of their kitchens lined with gallon jugs of water - thirty of them! They changed the front jug every day replacing it with fresh water. That way they always had a thirty-day supply of fresh drinking water available in case of disaster. Lists were also circulating among people I knew of items that should be kept in an emergency box.

I also believe that this was the time period when many of the survivalist groups, so prevalent today, had their start. One friend of mine actually dropped everything and went to Alaska to learn survivalist training. She had received the information, while in meditation, that this was where she was to go to learn survivalist techniques for that climate. These messages had to be very strong to cause so many people to give up their everyday life to prepare for the worst.

At this same time, I heard lots of stories of business men and women suddenly dropping everything and leaving their homes and families to go live in the wilderness. A minister of a church where I was the organist left his congregation to go live in Alaska under very different conditions than he had been used to.

An example of why these messages went out to so many people could be compared to weathermen who scan radar looking for storms that have the potential to form into tornadoes. When they spot a storm showing the potential to do this, what they call a supercell, they put out a "tornado alert" for that particular area. When they actually see the formation of the tornado, they put out a more urgent alert in the form of a "tornado warning."

I believe that the early messages received about coming destruction were in the form of "alerts" and as time went on these messages were more in the form of actual "warnings." This could have been due to the recycling or repeating that the messages were being subjected to, much the same way that gossip grows as it spreads.

Another example of alerts or warnings being given is that of an urgent situation which took place one night during the Cuban Missile Crisis, but in this case ordinary citizens were not warned, only members of the government.

My husband went to a meeting that night where the speaker was a U.S. Congressman. He told those at the meeting that our country might not survive the night. The situation was so serious that he intimated we had a 50/50 chance of living until morning. My husband told me this when he came home and it was a "heavy load" to contend with until daylight.

If this was the feeling that the entities in the fourth dimension had about our situation concerning cataclysms at that time, I can see why the message going out changed from an alert to an actual warning. At that time it was all they could do. During the Cuban Missile Crisis ordinary people did not get even that type of warning. Personally, I'd rather have a warning and have it change than not have any warning at all or have one too late to prepare for it. What an awful feeling that was.

It is difficult for us to understand that the possibility of destructive cataclysms has changed and that they are not going to happen the way that had first been thought. Perhaps this second movie will help you understand how first the alert was given and then the actual warning.

The title of the second movie of our double feature was "Drama in a Research Lab." I would like to describe this second feature film to you.

We are in a laboratory on a campus of a big university. In this laboratory we have many scientists at work on several different projects. One project has about ten people working on finding a better way to clean drinking water supplies than what is now being used.

Another group of about six people is working on a project with chemical compounds they are researching that will clean up radioactive spills in the environment. A third group is working on prevention of the radioactive spills. This is the smallest group with only four people working in this one.

The first group is working on a more direct relationship with everyday lives. The second is a generation removed, so to speak, from everyday lives and the third group is quite far removed bccause these spills are very rare. They don't happen very often and are widely separated by distance when they do. On the other hand, the chemical spills that the second group is working on are happening every day in separated areas of the world. The first group is dealing with something that is being used every minute of every day in all parts of the world.

The small group of four, working on preventing radioactive spills, discovers something that will affect what both the other two groups are working on. They discover a leakage of radioactive waste that is contaminating the soil underground at a site where many people are living. They alert the second group.

This second group then turns all its work force into ways to clean this leak up while it is still underground. At the same time they need to warn the first group working on clean drinking water as to what is happening. They tell them that the drinking water supplies are not contaminated yet but are in imminent danger of being contaminated very soon. They also tell them that precautions should be taken and plans put into place in the event that they can't clean up the spill in time or that the third group can't stop the leakage.

Therefore, the citizens are warned to bring in bottled water supplies to have on hand and to take precautions immediately against drinking any of the water from wells. They can continue to drink tap water because the city water supplies are not in the immediate area of danger, although if the leak is not stopped they could also become contaminated.

Meanwhile, the third group has been working frantically to stop the leak, even pulling some away from the first two projects to help. The second group has also been working very hard and they have come up with a way to neutralize the radioactive waste in the underground. The third group gets the leak stopped for the time being. It could happen again at another spot in the manufacturing plant where it took place because of the age of the machinery and the life span of the metals used in the machinery. Anyway, with the leak stopped for now and the chemical team able to neutralize the danger, the picture of the citizen's drinking water has changed. They are told that it is no long necessary to store bottled water and that it is safe to drink the well water, although it may be wise to boil it first just to be sure.

The picture that was first seen and the resulting precautions given to the citizens were both changed by the continuing research and development of radioactive spills and chemical neutralizers. This caused a totally "changed for the better" picture for the citizens. I hope this analogy will help explain how the picture of cataclysms could change.

At the end of our double feature movies we were given an added bonus in the way of a cartoon called, "Going Around in Circles." It reminded me of something that happened to me when I was a senior in high school.

My class went on its class trip to Washington, D.C. We were a very small class of twelve and even with a few teachers along were able to make the trip in three cars.

We were all very excited and as we entered the city the driver of the car I was in (who later became my husband) got caught in the rush hour traffic going around the Lincoln Memorial.

We spent a long time going around and around, trying

to work our way to the outer edge of the circle so we could get out of the mess and continue on. The other cars in our groups were by now already at our motel and were waiting for us and wondering what had happened. It was starting to grow dark and they were becoming concerned. We laughed and laughed about this later that night and still get a chuckle out of it today. We kept passing by the same spots over and over again, reading the same names on the street signs and going by the same bus stops with the same people waiting. It was very frustrating and very funny at the same time.

This is a good comparison to what happens in the cartoon, "Going Around in Circles." In this film we are being shown how we cycle around and around through the conglomerate mind. This fact adds to the volume of cataclysmic material being received. You see, our solar system is revolving faster than our galaxy.

To return for a moment to our large record of the whole Milky Way Galaxy, the same action as far as movement or spinning takes place in the big universe as it does for a record being played on Earth. I'm not sure how a CD works, but if you put an old 78, 45 or 33-1/3 rpm record on a turntable, the needle following a groove at the very center will make much faster revolutions than it does half way out or when it reaches the outer edge. This is what happens in our analogy of the record. The needle following the grooves in the tracks used by the planets in our solar system, near the center of the record, is revolving much faster than it will when it reaches the rest of the conglomerate mind as it moves towards the outer edge of the record.

We leave a certain area and make our revolution and come right back through the same area again and again because of the slower revolutions of the galaxy. For this reason, we will access certain areas or blocks of thought and pass through and move on into different areas or blocks of thought and eventually come back around again to the first area.

Let me describe this film. Instead of seeing us going around and around the Lincoln Memorial as happened on my class trip, I'd like you to picture an island with a car driving around it. Picture the groove in the CD that is allotted to planet Earth as the island and the needle as a car. You might have to stretch your imagination to visualize this picture, but please try.

Picture yourself driving in a circle around this populated island. You are in a car driving around the periphery of this island and passing through little towns on the way. Sometimes you come upon a much bigger city. You are in the city limits of this area for a longer time period than you are the small towns.

The energy is stronger in the city. This is due to heavier population which always tends to make for much stronger energy because differences of opinion cause friction, which sparks the energy. Therefore, the thoughts coming out of this city have a much better transmission system to follow – as far as sending out radio beams to be picked up on the car radio – than the thoughts in a small town have.

The same thing takes place in the Milky Way Galaxy. This is the holding place of the Creator's thoughts, and the the pole we have discussed could be pictured as a radio transmission tower. The records that left the strongest impressions on the pole are the ones sending out the strongest signals or radio transmission beams, which are then picked up by humans.

Eventually the car will pass through the city and move onto one small town after another until it comes to another large city. This city will have a different feeling, or mass consciousness, due to the events that led to this city being formed.

The same thing happens on the pole in the Creator's mind. Our entire solar system is moving through the galaxy at a faster speed than the Creator's thoughts are. So the whole solar system passes up the spot on the Creator's mind that sent out strong thoughts of cataclysms.

Therefore, even the people who are communicating past the pole of energy in the Creator's mind – in other words, communicating with the rings of energy, the higher planets – are still in touch with entities from other planets that are also passing through that strong thought-form of cataclysms. Remember, this thought-form was impressed onto the pole of energy by the inhabitants of Earth when the huge cataclysms took place in the past. The other planets can also tune into it becausc they watched while it happened. Many who were physically killed during the cataclysms have reincarnated onto these other planets since then. They are most likely the entities that people are in touch with when they contact cataclysm material from the rings of energy, which are higher than what is on the pole.

In our analogy of the car moving around the island, we should pretend that we have several grown children who each live in one of the different towns and cities we are passing through. We can be in touch with them for differing periods of time depending on the strength of the radio transmissions from each town or city. This would compare to a person having access to different former incarnations for certain periods of time as we move through the galaxy. The radio transmission beams go to all corners of the solar system from the Milky Way galaxy.

In our analogy of the island, people in the car would begin to lose transmission with the first city and would not pick up anything of much importance until they get within range of the next city's radio station.

In the universal mind, the next "city" may consist of a very strong thought-form about growing more nutritious food for people to eat. This is the thought-form that was present at the time of the Garden of Eden. The cycle has almost made it back around again to have Nature's Chemical Garden, a new nutritional garden to be discussed in a future book. The events that preceded the Garden of Eden were cataclysmic events forming the planet. Therefore, thoughts of cataclysms precede the thoughts of nutrition at the next strong spot on the pole of energy.

In our car analogy we have circled the island and have come up upon the first city again with the fights and riots, rapes and murders going on – cataclysmic events. We have moved through it to the next large city where thoughts of a different kind are present – thoughts of a better quality life perhaps. The car will keep cycling and as long as its speed remains the same, the time between the cities and towns and the time to make the full circle will remain the same.

The same thing is true as far as the solar system orbiting faster than the Creator's mind. It makes a full cycle every so many years right back through the same strong spots on the pole of energy in the Creator's mind. As long as the rate of speed of the solar system's orbit around the sun and the orbit of the Milky Way galaxy remains the same, the timing of the cycles will remain the same. Therefore, anyone beginning to make contact with the Creator's mind at the present time will not pick up the strong thoughts on cataclysms that people were picking up approximately ten to fifteen years ago. That cycle is past and we have not come back around to it again yet.

The rate of speed of the solar system as one big body and the rate of speed of the Milky Way galaxy will remain the same. The rate of speed of planet Earth's individual orbit is becoming slightly faster because of the addition of a new planet entering our solar system. This is causing Earth to move closer to the sun, thereby slightly increasing the speed of its orbit because it will have slightly less distance to travel each time. The solar system as a whole body, locked together into orbit around the sun, will not speed up. The planets are making the adjustment individually within the body of the solar system but the speed of the solar system as a whole body will remain as it was.

Have you ever tried to skip a stone across the surface of a lake? What fun that was when I was a kid! You had to find the perfectly shaped stone. It had to be flat and thin. My first attempts ended up with the stone sinking every time, but I finally got to be quite good at it and my sisters and I would have contests to see how many "skips" we could get to outdo each other. This same phenomenon happens with a radio signal also. It does the same thing that the flat stone does. It bounces across the air waves touching down at certain spots. (Come to think of it a tornado acts in much the same fashion.) This happens to the radio transmissions that the travelers in the car will then be able to pick up. Sometimes those in the car can still catch some of the transmissions about riots from the first large city, even though they have moved past it. It is a weaker signal but can still be heard.

The same thing takes place when Earth has moved past the strong imprint of cataclysmic activity. The travelers on the planet can pick up thoughts of cataclysms that have "skipped" out away from their original impression point. Due to this skipping of the radio signal, the really heavy thoughts can still be picked up many, many years after Earth has moved past that strong spot on the pole of memories.

When you add the phenomenon of the skipping of the radio signals to the fact that there have always been cataclysms and always will be, you have the effect of receiving signals about cataclysms all along your route – some weak, some strong.

Actually, the strong impressions that people were receiving ten to fifteen years ago were not from the real strong spot marking the original cataclysms. They were from one of the weaker spots to where this signal had skipped. You can see that there will always be predictions of cataclysms as long as people tap into their inner being.

There are many predictions of Earth changes that are visions of past cataclysms. At the same time there are still cataclysms to come. Some people see one, some see the other. To distinguish between the two when seen from the Earth plane is almost an impossibility. There will always be cataclysms until the planet is a dead entity as is our moon. Then will the activity cease.

Part II

Cataclysmic Behavior Cycles

Chapter Four: The Cycle of Buildup

Many of you have either been through personal upheavals in your life, are going through them now, or will be in the future. Perhaps you will understand these upsets more if you can understand the cycles of the large universe that impact the lives of humans. I would like to explain about these cycles in the big universe and perhaps you will be able to compare the events in your life to these cycles of the Creator.

The big bang was in reality a nuclear explosion which was the outcome of an insupportable buildup of pressure. What followed this nuclear explosion was another build up, explosion and cleansing, followed by another, ad infinitum. Between times of exploding and cleansing, the buildup period is one of great activity leading up to the explosion, and then the gentle calmness of the cleansing.

The buildup, explosion and cleansing form a cycle that repeats itself, in this case not in time, but in space because time and space are one and the same.

Therefore, instead of talking about a cycle coming back every so many years we must picture this cycle as spreading out from the center on a rhythmic pattern of space instead of time. This forms the galaxies and the solar systems and planets. As the cycles have not stopped and won't stop until the nuclear explosion has stopped, the forming of new universes, galaxies and solar systems will not stop until the explosion does. This will be when the force of the explosion is spent and the flow of subatomic particles is nonexistent. This, of course, means the stoppage of the nuclear explosion when it reaches its outer limits. There will be a period of nonactivity, different in each cycle, and then the buildup will begin again, which this time will be a drawing in of the energy to repeat the pattern of the original big bang again. This is the largest cycle. Perhaps you could picture this as a giant pair of bellows, or a giant lung breathing in and out.

The original big bang was a result of energy imploding or drawing in on itself. This was in a stationary spot because the universes were not formed yet. All the nuclear explosions henceforth will not be in a stationary position during the process of drawing in. The drawing in will come from all areas in a radius around the Creator. But it amounts to the same thing - an implosion. All will draw back into the tiny core at the center of the universe - the hydrogen atom that was left after the nuclear explosion. Then the whole process will take place again and again. The subsequent explosions will not be as potent and the resulting growth not as large on an ever-diminishing scale. This is the action that we refer to as a chain reaction. The further out it goes, the less powerful it is.

Meanwhile, within the big cycle are all the smaller cycles. Each galaxy, solar system and planet has its own cycles. There are cycles in space for all, even the planets. Earth is merely getting ready to make a move in the cycle of space (time) that has come according to the pattern of buildup, explosion and cleansing that is part of all space respiration cycles of the Creator.

We are living on the edge of the move which would be the final period of buildup just prior to the explosion, which will be the force that causes the movement of Earth into its new home, which will be as a cleansing. Then will come the period of rest, relaxation and peace that precedes the next buildup period of the cycle.

The thoughts that go along with this final period of buildup are that nothing can stop what is going to happen; nothing could stop the "explosion" from coming. Even though we don't know what the "explosion" consists of, the intensity must build to this point. Perhaps in your personal life this may be a divorce or the breakup of a meaningful relationship or quitting your job or resolving a family feud or any of the other things that cause upheavals in people's lives.

Nothing else in the world is as important as the release that the "explosion" will bring. It is as though we are blindly driven to it and we would "kill" or "murder" anyone who tried to interfere. We would "run over" anything that stood in our way. This is what the cells of our body go through just prior to the excitement of the "explosion." Nothing is important but that one goal, whatever it may be. The cells are running wild, in a frenzy. Excited, "devil may care" attitudes are present. The power of the release is held so important that absolutely nothing else matters.

An illustration of how this could affect a human would be anger. I sensed uncontrollable anger in a dream one night. It was not my anger; it was in another person just before he killed me in a previous life. I had been having nightmares for several years and that night I had been to one of my metaphysical classes. Something that was said, or a look or action by someone in the class, triggered a memory of this. Evidently my soul thought it was time I looked at it and understood it to help me resolve problems in my life. Therefore, the dream was in the form of a regression using symbols that I could understand.

Anger such as I felt in this other person is frightening. It is an all-consuming feeling of such disharmony with all that is good, that it gives off the essence of something so horrible as to be indescribable. It is uncontrollable. If you had this anger buried in the memory bank of your computer, think how sick you would be until you could pull it out and be rid of it.

This kind of anger is caused by the misuse of one's own psychic power. Perhaps this is unbeknownst consciously, but it sets up the circumstances that allow this uncontrollable, horrible feeling to take over. It is not being possessed by the "devil" or even another entity. It is being "possessed" by the evil in your own make up that you have actually taken pleasure in, thereby giving more power to this evil. It is this "evil" then that turns wild. In other words, it is energy running wild.

The person displaying this wants to stop, hates what he is doing and knows all the while he is doing it, that the real person that he is doesn't want to do it. He can actually feel it creeping into him; he knows it's coming and one part of himself watches fascinated but helpless, while this overtakes him, and it must be played out to release it. Whatever or whoever is in its path is destroyed. Yes, exactly like a tornado. The build up of energy forces is so strong and evil, that to contemplate it is to make one sick to the center of his being.

My dream showed me in my own vehicle (van) when the anger and subsequent beating I received from this man began. I knew someone in that vehicle was going to be beaten. I sensed this building up. Then I heard the screams as though coming from the vehicle behind me. (In dream interpretation vehicle means your body.) It was at this time that I left my body and was watching the beating and hearing the screams.

As the others led him away, I felt the horrible anger and evil sickness inside him, because by this time I had crossed completely over, although those leading him away didn't realize it then, nor did he. I subsequently felt more of his horror and sickness as he realized he had inadvertently killed me. I, meanwhile, loved him intensely as I had through all the previous beatings. Being on the other side by this time, I could understand his helplessness at the uncontrollable anger.

This buildup of anger is similar to what murderers and rapists and all evildoers on Earth go through when they commit their deeds of wrongdoing. Picture these people as being part of the cells of the Creator's body. In this case it is the cells of the Creator God over our particular solar system who are in the buildup period of the cycle just prior to the explosion. This is the period where they are in a frenzy and running wild with anticipation, ready to trample over everyone and everything that stands in the way of reaching their goal; the explosion or release of pressure.

The buildup period is what our universe, solar system and planet are all in at the same moment. The buildup after each cleansing is a period of great growth leading to an enlargement of the universe. The original big bang caused a flow of subatomic particles which eventually gathered together and formed into planets and stars, thereby allowing great growth to take place. This buildup period allowing for growth is the part of the cycle that allows unevolved souls the chance to reenter and possibly prove that they have achieved growth; that they have made progress. Every so often we come around to the point in the cycle where many of these souls must be given a chance to reenter the Earth plane. There are certain "rules and regulations," you might say, that have to be followed in relation to souls entering bodies and this is one of them. This cycle is for souls who are not highly evolved, to enter human forms for their growth experience on Earth. This is something that cannot be changed because these cycles affect all energy in the universe.

The entities entering bodies during the pre-explosion phase of the cycle are those who must be given the opportunity to return to Earth whether they are ready or not. Therefore, there are a lot of souls who have entered in the last twenty or thirty years or so who were not ready to enter, as far as soul evolvement, but who could not be stopped. They simply will either clean up their act or they will not survive to enter the cleansed Earth when we do get to the actual point of Earth making her move to the new orbit.

There are certain lessons that cannot be learned on the inner planes, or if they can it takes an unreasonably long time. Remember our discussion of horrible anger. This is what is in the soul of some of those entering now who really aren't ready but can't be stopped. In somany cases, a sickness of the soul such as this can only be healed while in a physical body on the Earth plane. Many times it cannot be overcome on the other side.

These souls cannot be denied the opportunity to prove or disprove whether or not they have learned their lessons. Therefore, during this cycle all the souls who normally wouldn't be considered ready to reenter must be given an opportunity. Sometimes they will refuse to reincarnate just yet, but for the most part they jump at the chance.

Consider it similar to the experience that another one of my brothers-in-law had. He had been studying for many years on a part time basis to become a CPA. He was married, and with a wife and two children to support was definitely in a hurry to receive his CPA license. He was tired of going to classes two or three nights a week at a college that was about an hour's drive from his home. He wanted to be able to go into business for himself in order to improve his family's lifestyle but he knew he had to pass a rigorous exam before receiving a license to practice.

Even though a professor who had him in his class advised him to wait before taking the test, he determined that he was ready to take the exam. He took it and failed. He settled down to more classes while working in another CPA's office to gain experience. He decided he was ready to take the exam again when the next opportunity came. He failed again and returned to yet more concentrated study. The next time the opportunity to take the exam came around he determined to skip it and wait for the following opportunity. He finally passed the exam. He had proven to himself and the world that he was ready to move onto the next stage in his chosen field, which was starting his own business.

We have been in this cycle for the last forty to fifty years, more or less, and these souls are incarnating that have not reached the level of soul growth they should have before entry. This cycle came about at the very same time that society was changing into that of a preschool society. The nurseries and preschools came about during World War II when women went to work in factories because the men were at war. When the war ended, women continued to work. This was partly because they liked having a chance to fulfill themselves creatively and partly because of what they felt was the need for two incomes in the family to improve their quality of life.

Several years ago I had an opportunity to gain first hand experience in preschools. I taught singing in the San Fernando Valley outside Los Angeles. I would go into a classroom with my guitar, sit on a little kid's chair with the children on the floor in front of me, and sing with them. Some weeks I worked with as many as five hundred children a week. It was an eye-opening experience. One thing I learned was the many different influencing factors that come to bear on children's behavior. Even though this behavior was influenced by events in the home, the time spent in the preschool had much to do with forming patterns of behavior for these children.

Something that could really upset everyone was change. Perhaps the school would decide to remodel the class rooms. Maybe they would just decide to change the classes around as far as the children or maybe a teacher would leave and a new one come in. Even a new child enrolled in the school would cause upsets. Children would have to learn not to take the new kid's toy or stuffed animal that they had brought with them from home for a feeling of security. Many fights between these very young children broke out over a scraggly looking stuffed bear.

I only saw each class for about twenty to thirty minutes a week, but I can say this: these nursery schools and preschools are one of the biggest influencing factors on what is happening with humanity today. Some of these children in preschools are the unevolved souls reentering that have to be allowed this opportunity. Patterns are set in these early years that in many cases can never be reversed.

For instance, a slightly older child would constantly take things from a smaller child and with this many children all wanting the teacher's attention, this older child would never be caught and corrected. When he grew up, taking things that belonged to others seems normal to him.

Some children were just naturally loud and boisterous and could scare others with their loud voices and scary faces. Some of them were caught at this and received considerable attention because of it, which caused it to become a way of behaving because the child saw it could get him attention. This, then, would carry over into his older years.

These situations that were not corrected in preschool contribute to much of their behavior as they grow older. One other thing would be lying. If they got away with it in preschool they most certainly carried this trait over into later life.

Many of these children weren't picked up from the preschools until very close to 6:00 p.m.. By the time they arrived home and had the evening meal it would closer to 7:00 p.m.. By the time the meal was eaten it was time for the children's bath and bed time. With the parents needing to be at work early the next morning, these children would be gotten out of bed, dressed, fed and left at the preschool well before 7:30 a.m. and most of them earlier than that.

You can see from all of this that the main part of these children's lives were spent at the preschool. There was very little time at home for any interaction with their mother and father. Quite often they were from single family homes with hardly any time for attention from just the one parent. When I realize how many children there are in preschools across the country and the world, it doesn't take long to figure out what has happened to our society.

There has never been a period of time in the history of the planet when so many children are committing crimes. This is taking place with younger and younger children as time goes on. We've all heard the stories of youngsters committing murder, sometimes just to see what it feels like. Just recently a six-year-old beat a four-week-old baby just to get his toy. This six-year-old was helped by eight-year-old twins. Is it possible that this same six-year-old got away with this kind of behavior in a preschool?

Three teenagers in my hometown were arrested for a series of robberies and had stockpiled an arsenal of weapons in a storage shed planning to murder a schoolmate just for the experience. They were all from "good" families, attended church, were good students, but still something was lacking somewhere in their make up. Is it possible that they are all some of these unevolved souls? If so, the level of consciousness simply is not there to prevent this kind of behavior and their abnormality was not discovered until it had reached this point. Even our methods in society of instilling this conscience through religious training and schooling failed to achieve the goal. There is the influence of TV in these teenagers lives to consider.

On a recent newscast, the parents of a sixteen-year-old boy were arrested and found guilty of being bad parents due to the things this sixteen-year-old had done. The parents had raised three other children with no problems. Isn't it possible that the sixteen-year-old is one of these unevolved souls? Should the parents be held responsible, or is our society truly where the responsibility really lies?

The problems come in society because of lack of understanding of all this. One of the problems is that souls who weren't ready to reincarnate but couldn't be prevented, reentered and had children. They do not have the necessary level of consciousness themselves to see wrong behavior and correct it.

In order to survive economically, these preschools can only afford to pay minimum wages which results in hiring workers sometimes without even a high school education, much less college degrees. In many cases these workers end up being some of these unevolved souls working in the preschools where the same unprepared souls, who are still coming in, are taken each day. Therefore, many of those working in the preschools do not have the level of evolutionary growth of the soul that would be able to see wrong behavior and correct it. In other words, what many of the kids in the nursery schools and preschools are doing is not considered wrong because the workers have not gained the level of conscience that is needed to guide these children.

Believe it or not, I actually could see a big behavioral difference in these children at the time of the full moon. Even the teachers were aware of this and looked towards this time of the month with dread. It was like being on a carousel or merry-go-round and at a certain point you come back around to where the music of the calliope is so loud you think your ear drums are going to split open. You begin to dread this part of the cycle of the merry-go-round that you're on, the same way the teachers would dread the full moon phase of each month.

Because most of these children were not old enough for the influence of their soul to have begun, they were living through instinctual behavior. Anyone or anything that operates on instinct is very responsive to the cycles of time in the larger universes. This is actually what distinguishes between the animal kingdom and the human kingdom, the ability to overcome the cycles of influence of nature.

Do you see the similarity of where the planet is at the moment, not only in its inhabitants but also in the physical move of planet Earth herself? The inhabitants are the cells of our Creator's body and are running wild in a frenzy ready to kill and maim anything that stands in their way of receiving the thrill of the "explosion." In many cases, the thrill of raping and killing is taking the place of the thrill of "explosion" that they (as cells of the Creator's body) can't experience until the cycle is complete.

The cycle has started that is the buildup to the final period just before the "explosion" that will propel our planet into its new orbit. The movement of subatomic particles in a wave of energy has begun in the outer edges of our solar system and is spreading slowly towards the planets. This will be a fairly slow buildup but will speed up and become quite fast when the time is right. The new energy waves will move all the planets at the same time.

Picture it as the buildup of molten lava at the base of a volcano which slowly rises until it explodes out of the top. Remember, the explosion period in all instances of the cycle causes new growth. The molten lava will enlarge land masses. The explosion out of the top of the volcano can be compared to the movement that will cause Earth to shift in her cycle as a new planet is forced into our solar system due to this growth cycle. Room must be made for it because it is caught in the magnetic pull of our sun. Once this move is made the new beginning must come right on top of it. I hope this lengthy explanation helps you understand that what humanity is going through is but part of the Creator's cycles and is a natural, normal process.

Once the shift is made there will be the period of cleansing, remember, and those who do not clean up their attitudes will be removed and reconditioned on the inner planes until the time of the next cycle when we will not be able to keep them from entering again.

These cycles should be studied and charted and periods of human behavior will clearly come to light and the amazing precision of these cycles will be apparent.

Chapter Five:
The Cycle of Explosion

There are two different causes of cataclysmic activity on Earth. One is caused by scientific reactions to events in the large universe (physically caused cataclysms) and the other by emotional upsets in the humans living on Earth (emotionally caused cataclysms). The physically caused cataclysms happened at the time of Atlantis. This is not to say that there won't still be physically caused cataclysms on Earth, but not of the nature of the polar shift that took place at the time of Atlantis. The emotionally caused cataclysms, or upheavals, are happening now.

The same cycles of buildup, explosion and release that take place in the large universe also manifest on the Earth plane in the behavior of humans as discussed in the last chapter. For instance, the area around Los Angeles is an area that is so crowded that every inch of available land is in use. Whatever is taking place on the surface in these communities where it's crowded, where people fight and argue, will eventually cause them to released. When it is released it goes into the earth. It actually sinks into the earth.

As you know, when people do clearing exercises, they send the bad energy that they're clearing from people right down into the earth. Also, when people cleanse their physical bodies of poisons by purging through diets, fasts and other means, this human waste is also sent into the ground through the sewer systems in our land. That's where it's sent to get rid of it, and that is the energy that is causing the motion down there. Earth responds the same as our solar plexus that quivers when bad things are happening. That's what happens to the planet as a result of negative emotions. It starts the motion that will eventually cause the slippage.

The emotions of those living in the areas where a quake occurs are causing the upheavals because of the negative energy of these emotions. The emotions are caused by a lack of spiritual consciousness; a lack of personal growth in most of the population areas that are hardest hit. Some of the worst earthquakes have been in areas where the consciousness level has not been raised (here again, unevolved souls who reentered) and there have been thousands of people killed at one time. You see this to some extent in all earthquake areas.

A week after the devastating earthquake took place on January 14, 1994, a program about earthquakes was on TV. During the program, a map of the world was shown and it was pointed out that there was a direct correlation between really destructive earthquakes and large popu-populated areas.

What you're going to find is when this negative energy goes into the earth, it's going to be just exactly like the human body. It's going to affect the weakest part of the system it's being sent to. In the case of one area it would be floods that were caused, in another area it would be fires. All this activity that is going on in nature is purging. Earth is trying to purge itself of these negative vibrations. These are the cataclysms that are taking place that have been forecast. This is what is meant by the fact that the cataclysms will be in men's minds. There actually will be both physical and mental cataclysms. One will cause the other. The cataclysms, the bad attitudes, the lack of consciousness, are in people's minds. Their negativity is going into the earth causing the physical manifestations of the cataclysms.

While living in California I experienced floods and saw houses sliding down hillsides. I was there when the Rodney King beating took place, actually lived a short distance from where the trial was held. I experienced the riots afterwards. I was teaching in one of the preschools during the worst day of the rioting. All the parents were called and those who were able came to get their children and take them home. The children whose parents couldn't be reached or couldn't get away to get them, were all gathered into one room of the school. The blinds were closed, the doors locked. All the teachers were there also, including some who shouldn't have been because they were hysterical (unevolved souls who reentered?) and upsetting the children. I sat in the middle of the room and sang with the kids and teachers while the rioting went on a few blocks from the school.

I also experienced one of more devastating wildfires the state had. Standing in the front door of my living room I could look across the way to a hillside that was burning closer every minute. The smoke filled the air. Many people had packed important belongings in their cars and were ready to leave at a moment's notice.

I was there when the murders of Nicole Simpson and Ronald Goldman took place; I was actually traveling the opposite direction on the same freeway at the time when O. J. Simpson's white Bronco, followed by the police cars, were traveling on it. My son and I were heading for the airport and heard it on the radio. Our exit ramp came up and we left the freeway at the same time the cavalcade was going by in the opposite direction.

But the worst of all things I experienced was the earthquake on January 17, 1994. I was awakened at 4:30 in the morning with the bed going up and down like a bucking bronco. My niece from Phoenix, her eight-year-old daughter and her eighty-two-year-old grandmother were visiting us. My son was sleeping on the couch in the living room and was thrown onto the floor. I was concerned about the grandmother in the next room sleeping right beside a big window. It was difficult to stand up and walk straight to get into the bedroom next to me where she was. We had a wheelchair for her, but it was out in the garage. She had difficulty walking but we finally got her into an desk chair with wheels and pushed her into the living room where the rest of us had gathered.

Miraculously, when we turned on the TV it was still working, but only for a few seconds. A technician had been shoved in front of the camera because there was no one else there and he was very "rattled" and upset. He tried to relay what was happening but things were shaking all around him and then we lost power. My son went out to his car in the garage and turned the radio on until we finally found a radio in the house that had batteries in it. Because it was growing cold we all covered up with blankets and sat huddled in the living room listening to it. As each report came in the enormity of what happened began to dawn on us. Actually there was no damage done to our house. Some friends of ours in the next town didn't fare so well. My son knew the parents were out of town and the four children were alone so he went to check on them. The oldest was eighteen and the youngest four. Their house was a disaster but no one was hurt.

The effects of this earthquake lasted for months. To begin with there were the aftershocks, in themselves as scary as another earthquake. Driving around the San Fernando Valley it looked like a war zone. People were living on their front lawns, afraid to be inside their houses in case of another strong quake, or in National Guard tents set up in city parks. It was cold and rainy. Those whose houses were okay didn't have heat or electricity. Food and water were scarce. Most businesses were shut.

I had been going into homes teaching music lessons and after a couple of weeks I decided to start teaching again. One of my students had lost her piano totally – the brick chimney had fallen on it and smashed it. Another student had a large aquarium fall onto the piano spilling water and fish all over the piano and smashing the bench. Another one's house was very close to the epicenter and after seeing it I was afraid to go into house every time I went back there to give lessons. The walls were cracked everywhere and all the tiles had fallen off the wall over the fireplace.

The church in Hollywood where I was the organist was pretty badly damaged structurally and also lost some stained glass windows.

One preschool where I worked consisted of two separate buildings, one around the corner from the other on a different street. They were very close – it probably took less than a minute to walk from one to the other. One building was not even damaged; the other was completely destroyed. A few hours later and it would have been filled with children and teachers.

The office where my son worked was a disaster area. He and his coworkers were only allowed back in for a few minutes to get important papers. Computers were everywhere, lights were down on the floor, bookcases overturned and there was much damage to the walls and ceilings. It was also within a short distance from the epicenter. I am so thankful that no one was in the office at the time. There were no people in any of the offices because it was so early in the morning. Here again, a few hours later and the results of that earthquake would have been horrible. All in all, it was an experience I will never forget.

Many of the cataclysms that have happened, for instance this LA earthquake, present people with an opportunity to awaken to the spiritual growth that they need to have (here again, an opportunity for unevolved souls?) It will affect some people in this manner, but not the majority. They're not ready, not in this incarnation (not evolved enough – should have waited?) Perhaps only fifteen percent (15%) of the people who suffered extreme damage are going to start searching their souls (but this will be that many more souls who gain a consciousness level needed to keep progressing spiritually).

They will start searching their souls as to why this happened to them, why they made a decision three years ago to move into that particular house, for instance. To start this search is the best way to open people up to spiritual growth. The same thing will be true of any of the cataclysms. Those people who went through the floods, those people who lost their homes in the fires, they're all starting that mental search, which is where it must start for the growth to take place. It is a self-perpetuating process. Remember, in the cycles of the large universe, the period of buildup precedes the period of explosion which causes growth to take place. People actually cause the cataclysms that cause their growth. It is a self-repeating cycle.

When people are buried in the earth those vibrations are buried with them. This is why people should consider cremation rather than actually burying the physical body in the depths of the earth. Those negative vibrations are there. If they could see the vibrations over cemeteries where the bodies of these souls are put in the ground and left to rot, they would not bury the dead anymore. The negativity doesn't rot. It goes deeper into the ground and spreads through it.

A medical procedure called an MRI can trace a substance through the body to see if there's a blockage. The same thing could be done with this negativity that is in the earth. This would help with predicting where the next earthquake is going to take place. This could also be done for floods and droughts. Please realize that this isn't anything that those in the higher realms can control. This must be understood – they can't control it. Those of us on the planet are the only ones that can control our own environment. There are some new age gurus who say that space people can sit up above in their flying saucers and tie down a bridge abutment, etc. This is nonsense! It doesn't work that way.

Events are instigated by thoughts. People need to wake up and see the whole, big picture . . . and it's very difficult for one individual to see that whole picture. This is why they need to consider the fact that they are individual bits of one mind. They may think one little individual can't cause a calamity, but that individual, with all the other individuals, makes a big, powerful mind – negative mass consciousness.

Cataclysms occurring over the last fifteen to twenty years should be tied together on a map. This would include all the major catastrophes, no matter whether it involved hundreds of people or as few as fifty or even less, anything that would be considered a catastrophe. If pins were put in the map as to where each event took place, this would alert people. Different colored pins could be used for the different numbers of people who were affected and for different categories of catastrophes. It would start them thinking about what their thought processes do to the body of the planet.

People in California will bounce right back but, as said, only about fifteen percent of those badly affected will do some soul searching, but it's not enough. When you look at the segments of populations that have the most children, you can see that it's a self-feeding thing that perpetuates the lack of consciousness. They bring their children up with the same lack of consciousness that they have. How is it ever going to get turned around? Even those beings in the higher realms wish they knew how! They keep trying everything that they can possibly think of and they still don't know how to change it around.

Because thoughts are things, the recent Kobe earthquake in Japan, which came a year to the date after the Los Angeles earthquake of January 17, 1994, is a direct outcome of the LA quake. Actually, the strong imprint of the LA earthquake left such strong impressions on the body of the planet that it responded with an earthquake on a small cycle of one year in a location very near the first one.

The point I'm trying to make is, here again, it's all caused by humanity itself. People don't have any idea of how much they are responsible for it all! The earthquakes are a reaction of the planet itself to the calamitous emotions being felt by the inhabitants. This is so very difficult for people to accept.

Perhaps the following analogy will help people understand: Please picture Earth as a mother dog with a large litter of, say, sixteen puppies. These puppies nurse from their mother, but they also crawl all over her playing. They are all in a cardboard box which limits the area the puppies have to roam. They are more or less forced to live their life on top of their mother's body.

As the puppies grow larger, they also come into the animal instinct in very strong degrees and these animal instincts bring on fights. Upheaval is prevalent at all times within this box. The mother tries at first by giving warnings and little nips and perhaps growls to control her brood, but it soon becomes impossible to struggle against them.

She resigns herself to this sort of activity on her body and slowly but surely the vibrations of the fighting dogs begin to wear away at her vital organs. It first affects the muscles of her body which causes her to not be able to move out of the way of the puppies into the corner of the box. She is slowly becoming paralyzed.

As the puppies continue to fight they are also continuing to nurse and get their sustenance from their mother. The mother's muscles become paralyzed, partly from the vibrations that are constantly being sent into her body through their nursing and partly from their animal instincts when fighting. She can no longer consume enough food or digest it in order to provide the milk to nurse the puppies.

The milk processing plant of her digestive system is the first to shut down. The other organs soon follow suit, but then a strange event will take place. There will come a period of time when the animal instinct body of the mother will take over and rise up as energy – unbelievable energy – in an effort to save her own life. She will find the strength to stand up and shake her body and thus shake loose the puppies and scatter them against the sides of the box. This is an inborn instinctual happening that actually would take place with humans who were being destroyed in the same manner (if this was ever to take place).

At times this instinct will show up in humans as they prepare to die. In most cases it does not, due partly to drugs that are used and also to the fact that human beings are leaving their instincts behind as they evolve. This is one of the instincts that survived the transition from the animal stage to the human stage. It has slowly been bred out of the race because it was not needed as strongly in humans as it was in the animal kingdom.

This mother has now reestablished her power in the cardboard box. The puppies settle down, learn to eat other nourishment that is given them and do not continue to feed off their mother. This heals her digestive system, allowing nutrients from her food to heal the other organs of her body. If the puppies are not removed from the box the whole set of circumstances will take place over and over again. It will continue until the puppies stop fighting and calm their vibrations to a peaceful, loving coopera-

cooperation with each other and their mother.

Now, let me relate this to communities affected not only by this earthquake, but also by all disasters. People can readily see the analogy of humans living on top of "Mother Earth." When they fight and bring forth their animal instincts they affect the functioning, the harmonious pre-planned functioning, of "Mother Earth" as being the one able to provide sustenance. Please keep in mind that the human body is made up of material of the earth. As our bodies are sensitive seismic instruments of stress, so also is the Earth from which these bodies were fashioned. They are made up of exactly the same material!

The first thing affected in the mother dog is the muscles. The way this compares to the planet is that the plates allow movement of the continents, just as the muscles of the dog allow movement of the dog. The muscles in the dog become weak and stop her ability to process any of the bodily functions. So also does the stoppage of the movement of the plates stop other processes in the Earth. These other processes that are stopped are the cause of floods, fires, volcanoes, tornadoes, snowstorms, lightning and hurricanes. All the other natural disasters that take place are the result of the stoppage of the movement of the plates. This is caused by the animal instinct emotional upheavals that people living on "Mother Earth's" body are sending into her body.

When things get to the point of being no longer bearable, the natural instincts of the mother will surge forth. She will stand up and shake to remove the burden of the parasites living on her body and taking their sustenance from her nourishment. Then there will be a healing process until the whole thing happens again.

It must be realized that all the crimes, violence, riots, rapes and murders result in emotional upheaval. This emotional upheaval in the Los Angeles area is not only the cause of the earthquakes and floods and fires in that immediate vicinity. It is also the direct cause of the non-functioning of other parts of the planet. This would include the floods in the midwest and the snow storms in the east. Of course, the same emotional upheaval in other cities around the world would all add to the non-functioning of the planet, each in different areas.

The other thing to keep in mind is that the mother dog is not going to stand up on just two of her legs. She will have to find her balance by also standing on the other two legs. This will be almost a simultaneous happening and is the major cause of the aftershocks. What is going on in the Los Angeles area now is activity on three different fault lines. The emotional turmoil and resulting vibrations that have been sent into the earth at that point have been spread over all sections under the city. Therefore, more than one fault line has been activated in the urgency of Earth to shake off these unbearable negative vibrations that have been coming down into her sensitive body from the people living on her in those areas.

So not only are there counter-balancing earthquakes in the form of aftershocks in the immediate vicinity, as the mother dog puts down each of her four legs and stands up, but as she shakes she causes motion that rocks the whole box (the whole Earth sphere). If she is on one side of the box the puppies will rush to the other and cause upheaval there. People themselves will not rush to the other side of the box (planet), but the vibrations from their animal instincts that are being released will. In some cases this will cause, in the opposite side of the box (planet), the sensitive Earth to erupt in a volcano or perhaps another earthquake in a attempt to find balance. If the vibrations are shaken outside the box (planet); if the shaking of the mother (Earth) is strong enough to send the vibrations up and out of the immediate area of the box (planet), then these negative vibrations will join the streams of atmosphere causing disruptions to the weather patterns. As a result there will be unusually heavy activity in increased rain and snow and windstorms, which will in turn affect not only the planet some more, but also the oceans. For the most part, these storms are signs of healing that are going on within the other organs of the mother's body.

It is quite difficult to ask people to see this picture and stay calm when all that is happening to them has made them depend more heavily than ever on their animal instincts of survival. Perhaps it will help with understanding in some areas, but it most likely will not help enough to stop the advancing severe natural disasters that are ahead of the planet.

Chapter Six:
The Cleansing Cycle

People must begin to think of their planet as a living entity. Earth reacts to those living on it, and when their lives are in emotional upheaval,Earth responds with severe natural disasters. The whole picture of cataclysmic events is totally controllable by humanity itself. Therefore, everyone needs to raise their consciousness to prevent, or at least lessen, these disasters.

The cleansing of the planet and of the people living on it must both come about at the same time. To do otherwise could be compared to changing the bedding on your bed so it was nice and clean and sweet smelling and then crawling into it straight from a mud wrestling contest without benefit of a shower. In either case, the cleansing will be done by raising the vibratory field. This is what is taking place with the planet as it moves into its new orbit. It is moving into the fourth dimension, slightly higher and slightly closer to the sun into a clean field of space.

On the other hand, people must raise their consciousness first to then be able to raise their vibratory field. Raising the vibrations of the physical body will come in response to mental thought which is similar to moving from the third grade to the fourth.

To understand more about why this transition to fourth dimensional thinking must come about, I'd like you to picture a grade school. Let's focus on those in the third grade. When school is out, they play all summer and when they return to school in the fall and enter fourth grade they are still of third grade educational level. They will not be true fourth graders until near the end of the school year. Somewhere in the middle of the school year or towards the end of it, the child will begin a slow transformation into fourth grade mentality which he must reach or he will not be passed to fifth.

The same holds true of the planets. Each planet will remain the consciousness level of the previous planet until enough of the inhabitants have raised their consciousness to reach the actual level of the planet they are on.

On the first planet, life was simply in the form of air. On the second planet, life was in the form of clouds of atmosphere, the air forming into groups. On planet one, the air was a thinking mass of air, the Creator's thoughts. On planet two, these thoughts separated into those of like mind forming groups or clouds. On planet three, the clouds built up into large storms with all clouds forming back into one huge angry storm that moved, in the form of lightning, to planet four (Earth). When it hit planet Earth, it exploded into millions of sparks.

Planets 1-2-3 were the conception, incubation and gestation periods for humanity. They were manifested in physical reality in conglomerate (unseparated) form. But on each of these planets the forwarding impetus to the next planet was reached in thought before the move was made. In other words, on planet one the thinking mass of air played around on the planet until it decided there must be more to explore. This thought was singular because the mass of air was singular. It immediately rose up into the atmosphere around the planet, the aura, and moved immediately to planet number two. The aura around each planet is a collecting point. You might compare it to a holding tank that collects so much water before beginning an automatic releasing of the water. On planet two they separated into groups or clouds of like mind but still in communication with each other. They stayed here until one group had the thought that there must be more and immediately moved up to the aura or holding tank for planet two.

When the other groups discovered what the first group had done, where they had gone, they followed one by one. When all clouds had risen into the upper atmosphere of the planet, they moved as one body to planet three. Please keep in mind that we are talking about thought embodied in matter, but this matter is the atmosphere.

On planet three, they soon found that each group or cloud was growing larger. There were fights breaking out and arguments as to which group had better ideas or more valid ideas that needed to be experienced. The only form these thoughts had to express this anger in was the clouds which were building into huge, fierce storms. Instead of one group moving to the aura or holding tank, they all formed into one big free-for-all fight or storm cloud. They knew they couldn't stay on this planet any longer, there had to be more. That thought launched the storm onto the aura and immediately, because they were all one body again, made the move to the fourth planet. By the time this storm reached the fourth planet there were fierce lightning bolts zapping around. At one point it all ignited into one huge lightning bolt and this shattered the cloud of energy into millions of sparks as it landed on the fourth planet.

Now, planet Earth had physical forms manifested in matter on it. It had a mineral kingdom, a botanical kingdom and an animal kingdom. These thinking sparks tried to enter the animal forms, found they couldn't control them and waited until the forms were refined through evolution that was speeded up by divine help. They entered the animal forms again and evolution continued to refine this part of the animal kingdom that had broken off from the main part. This is humanity.

Instead of being in small groups on this planet, in the beginning all were separate, totally apart from the others. This was a level of mind that developed at this stage, or the first measurement of thought that can be determined, that the brain was capable of: first dimensional thought. When you take the measurements of something, you are determining its dimensions. Thus the use of the term dimension of thought as it truly is a measurement of thought.

When the sparks begin to roam in pairs, male and female, and form family units, they entered the second measurable enlargement of mental thought: the second dimension. When the groups of families began grouping together into tribes or clans and needing to have leaders, they moved into the third level of mental capacity, the third measurable level of thought or the third dimension of thought.

As on the previous planets, some of the groups on the fourth planet begin thinking that if they have come this far there must be more to experience and began separating themselves from the groups and stretching their minds, which elongated the measurement of thought until it finally reached a separate level of thought unlike anything that the groups were still caught up in and thinking. This is fourth dimensional thought. This moved these individuals mentally to the aura of planet four, the holding tank of the planet where all must get before they can join back into one big mass of thought or one big cloud and move to planet five.

Therefore, humanity is split between third and fourth dimensional thought on the fourth planet. Those into true fourth dimensional thought can only be in the fourth dimension in their mind until everyone on this fourth planet reaches this level of thought. To do this they must break out of group thought that organized religion holds them in or that family ties, through their physical evolution roots, hold them bound to.

You see, the whole planet should be in fourth dimensional thought because it is the fourth planet, but it doesn't work this way as explained about the grades in school and a child staying third grade mind level when it enters fourth grade. Do you see the analogy?

Planet Earth is truly a classroom experience, as are all the physically manifested planets. Even though the life forms on the other planets are not in what we term physical manifestation, they are in their own form of physical manifestation which happens to be of a higher vibratory rate than Earth.

To get back to our explanation, there was a physical location for first dimension thought, the first planet, and a physical location for second dimensional thinking, the second planet, and also a physical location for third dimensional thought, the third planet. By the time they splintered on planet four the first dimension was dropped off. It was no longer available because it had been one large thought-form and once it splintered this was no longer the case. So when humanity entered the fourth planet, it had already lost first dimensional thought capability.

Dimensions of thought will follow atmospheric conditions. Second dimensional thought that had taken place on the second planet, that of clouds dividing into groups of like mind, took place on planet Earth as individual families. This then disappeared simply because of the need to form into groups for protection against animals. Therefore, individual family instincts of looking out just for themselves were left behind because they would soon perish at the hands of the big animals if they did not form into clans or groups of families for protection. When they did this, they became the tribes that moved to tribal thought and second dimensional thought, left over from their stay on planet two, was left behind. Now they were all in third dimensional thought. The third planet was where the groups of like mind, individual clouds, all formed into one big storm cloud.

This is the stage humanity is still in and has been in since the days of the big animals, except for the few who are beginning to break away from established thought and move out on their own as unique individuals. This would be by getting back to the stage they entered the Earth plane in when the lightning bolt, that had formed out of the storm cloud on planet three, shattered into billions of sparks. These were individual thought-forms of the Creator before they were caught in the physical bodies which forced them to regress into group thinking.

Regardless, third dimensional thought had a physical location, the third planet, and fourth dimensional thinking has a physical location, the fourth planet, but it is and can only be in the mind, in people's thoughts, that they can experience the true beauty of what the planet should be by now. They cannot experience it as a physical reality until all of humanity enters fourth dimensional thinking.

This will be when people come out of the cloud (the big storm cloud of the third planet thinking capability) that they are under at this time. They will be of a much more mature nature. The eons that went into making up this cloud (third dimensional thought) have been a classroom experience for us. We need to graduate out of this cloud or "fog age," if you will, as we did the dark ages, the stone age, etc.

The mental chaos people are going through causes confusion and fear. This will cloud and slow the thinking processes of people and prevent them from accessing the higher realms. This cloud has to be shattered as though it were made of brittle material. Remember, thought capability of humans will follow atmospheric conditions. Therefore, because our weather patterns are going to become much more erratic, so also will our thinking abilities become erratic and therefore our economy will worsen. When humans are faced with trials, they grow the most. It must come to this.

This process of atmospheric conditions raising the thinking capacity of humans is ever ongoing. The changes in the energy waves around the planet will be what causes the changes in the brain wave patterns of humans. The brain was restructured by the use of atmospheric conditions at the time of Atlantis when the Creator used help in the form of electricity from meteor showers and lightning bolts from the higher planets. This restructuring of the brain is what allows a lifting up in men's minds/thoughts to the higher realms to take place.

This lifting up happens at night and they are returned the next morning to pursue their missions: to start implementing what they learned. Because two hours on the inner planes equals twenty-four hours of real time, it wouldn't take long for a person to accumulate the equivalent of years of training on the inner planes. This actually is the very beginning of moving from third dimensional consciousness to fourth.

Because more and more people are entering the new consciousness, that is, moving from third dimensional thinking to fourth, the emotionally caused cataclysms, or upheavals, are not going to happen as violently as previously feared. They are being dispersed over a longer time span, which lessens the impact they could have had. This does not mean that all should give up working on their attitudes, by no means. It is a grace period for more to become awakened. Nothing remains as it was. All plans are and should be open to constant revision. Not to be open to change is counter to evolution. Some of the changes are for good, others not. It is a choice of how energy is used. This choice (free will) has always had the power to change not only the universe at large, but also each person's individual universe.

Referring back to the cycles of buildup, explosion and cleansing in the large universe, the cycle of cleansing is followed each time by another buildup period which is growth. Actually, at times these cycles overlap each other as far as how they affect humans. This is due to the fact that some are cycles of the large universe, some are cycles of our galaxy and some are cycles of our planet itself. The floods, quakes, fires and volcanoes, all the spectacles of nature, are part of the cleansing cycle of the planet that must come about to allow the buildup cycle of growth for humans, which is the movement from third dimensional consciousness to fourth, to take place. These cycles of cleansing take place periodically when the underground currents become saturated with the refuse that people wash into it. Here again, we instigate our own growth process.

I cannot pinpoint the exact start of this present cycle of cleansing. The closest I can come is it was sometime between the 1940's and the 1980's. I can remember, when I was a kid back in Ohio, every summer our whole family would pack up the tents and camping equipment and head for Lake Erie. My first memories of this were of a beautiful, clean beach and a well kept campground. There were a few cabins but most everyone camped in tents. This went on for perhaps six or seven years before we noticed any change in the lake. We began to notice how dirty it looked with all the debris floating in it.

One year we got our tent all set up and headed for the beach ready to have a great time. The first one of my sisters to go into the water turned right around and came back out as fast as she had gone in. She said, "It stinks." My mother waded out a little way and came back and said that none of us were to go into the water again. There was human fecal matter floating all over in it along with many other things. We were sickened and saddened by this turn of events and after a couple of days we packed up and went home. Needless to say, that was our last camping trip to Lake Erie.

I revisited the area in the 1980's and the water was much cleaner than the last time we had been there to camp many years ago. Actually, the beach is no longer there, nor is the campground. The water had moved in and torn down the wooden steps that led to the beach and ate up the beach itself. The broken steps ended in the water where previously there had been a very wide area of sand between the steps and the water.

An exercise that everyone could use for cleansing the planet would be as follows. If you have occasion to visit a lake that several years ago was sparkling clean and is now polluted and dirty, the best way to help the situation would be to hold a mental picture and thought of the lake as it used to be. This should be done once a day, preferably at bedtime. Send this picture and thought out into the ethers. Pick one or two specific areas in your locality and concentrate on them for a period of two months, then move onto another area of concern. You may just find that you will start receiving mental images and thoughts back as to how these areas could be helped. This just might motivate you to take on a certain project of clean up as your own spiritual mission to the planet and her occupants. Can you understand how this could help the planet if everyone in the world took part in a cleansing of the planet in this manner?

The furious storms that will cleanse our atmosphere will begin when the area under the ground is cleansed. The garbage under the ground must eventually come up through the atmosphere so there is no point in cleansing the atmosphere until this takes place. At that time the atmosphere close to the planet's surface will be cleansed. The cause of these furious storms will be the move of our planet into its new orbit.

We are in a large universe with many suns. These many suns have many planets that have been attracted into their orbit. When the rhythm controlling these orbits is thrown awry, these satellites then change their position ever so slightly and reduce or increase the timing of their orbit. Some will move slightly. Some will decrease their revolutions while others will increase them. Earth, itself, will increase her spinning which will decrease the length of time it takes to make its revolutions around the sun.

Our time is already speeding up; our days are becoming ever perceptibly shorter. This is already starting to throw off the timing of every conceivable function on the Earth plane. People are noticing seasonal changes: tornadoes happening all year long, longer hurricane seasons with one huge hurricane after another, flooding, etc. Even animals are being affected with differences in their shedding cycles, etc.

Panic will begin to set in when the speedup of time starts to be felt more strongly on Earth, because our time keeping devices will need to have ever increasing adjustments to keep pace. The lack of understanding as to just what is happening to our planet will cause chaos. This chaos then will, of course, affect the internal part of our planet, which will then be expressed by increased earthquakes and volcanoes. Furious electrical storms will be increasingly common because of the changes in position of Earth's orbit. These changes will put her in an untouched field of magnetism that will respond to the friction of a spinning planet by throwing out sparks which will be as electrical storms.

This is the process that will cleanse the planet not only outwardly in its atmosphere, but also inwardly, because these bolts of "cold lightning" will penetrate deeply. It is referred to as "cold lightning" because it is different from our normal lightning. It is being caused by a virgin territory of space being impregnated by a "foreign object": Earth. When this happens, the results are a phenomenon of great interest and excitement for all in the immediate area, especially those entities on planet ten (the experimental planet). They are the ones who have planned this cleansing in conjunction with events that were going to happen anyway. This "unknown" territory, this time-space block of suspended animation, will be brought to life by the movement into it of planet Earth. This is the reverse action of what started life on our planet and this is the tying back in on itself of everything in the universe. This will actually start the process of withdrawal of the oversoul from our planet to continue the journey on through our solar system.

Those entities on planet ten are serving in our galaxy on a special mission to raise Earth's vibration and to save her from self-destruction, not necessarily through war but through neglect of the planet itself. They also have direct knowledge of what will happen to a planet if it is treated as planet Earth is being treated for too long a time. Therefore, instead of progressing onward in their own spiritual climb, they have elected to serve in this galaxy for a period of time to help. If they did not feel brotherly love for us, they would not stop their own progress to help us.

The bad, very destructive cataclysms have already happened. There will still be natural cataclysms and some of these will be destructive, but it will not come to the point where we are in any danger of being destroyed. For one thing, we're much more advanced technologically because many people have already been lifted up mentally into the inner planes at night and trained and returned to bring in this new technology. Therefore, we can overcome and protect ourselves from many of the natural happenings. Some thoughts have been given on ways to do this, such as information on water disbursement stations and a new solar energy project (both of which will be discussed in later books). These projects fall into the category of protection against cataclysmic events. Another way of protecting ourselves would be by understanding and using new thoughts on earthquake prediction information as given at the end of this chapter.

While I was living in California, my central nervous system was upset all the time. Being similar to a finely tuned seismic instrument, I sensed the turmoil in the earth under me in California. It was always like being on the top of a bowl of jello. That was the way my nervous system felt. There was an earthquake that took place June 28, 1991. The day before this earthquake there had been a mysterious jolt. I felt this jolt as though it ran horizontally the length of the apartment building. It traveled as a straight line and seemed to be just below the second floor windows.

This mysterious jolt was a different kind of planetary movement, one that could not be detected by instruments now in use. Earth herself hit a "pocket" of vacuum left from the original explosion. There have been several mysterious jolts recently not only in our country, but also around the world. Picture it as a bowl of water that is traveling along a conveyor belt. Suddenly, for a moment, the belt floats free of the drag that is on it. This causes a slopping of the water inside the bowl. This is what happens inside the planet. The molten lava in the center of the planet slops up against the sides and causes all sorts of problems. This is a natural force of nature.

As Earth shifts into uncharted seas of eternity, this will happen more often. Instruments above the planet are the only ones that can measure these "bumps." These jolts are not as dangerous as earthquakes themselves. It is just a momentary absence of normal gravity pull that causes a slight bump, but they will serve as warnings of movement within the planet. The next time a "mystery jolt" is felt, it should be noted that within twenty-four hours the planet will also move from the inside. They are related.

A worldwide map of all quake activity needs to be kept and monitored at all times and studied, with locations and dates compared to find the pattern of counter-balancing.

Those who are doing earthquake prediction still are not looking over the fracturing pattern of the whole surface of the planet. They need to keep in mind that what

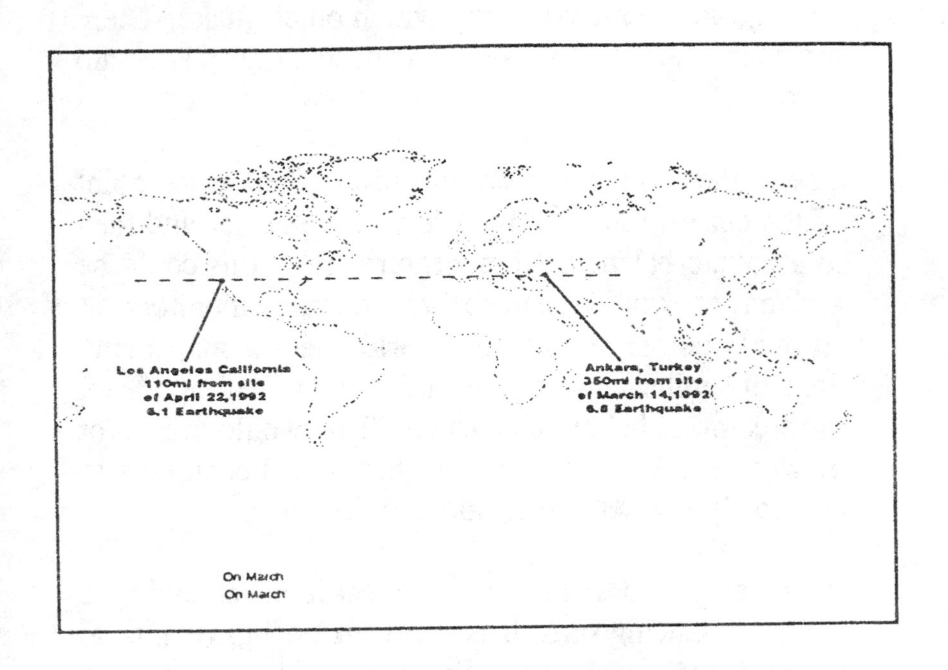

On March 14, 1992, at a location 350 miles from Ankara, Turkey there was a 6.8 earthquake. On April 22, 1992 at a location 110 miles from Los Angeles, California there was a 6.1 earthquake. — Illustration by Neil Hickox

occurs in one area will always be counterbalanced in another area. It is not as simple as it may sound to find what quake counterbalanced which other quake. There are many factors involved that need to be taken into account.

One of these is the temperature of the shell at the point of the quake, followed by a line to the center and then to a counterbalancing temperature. How this could be arrived at would require temperature monitors at strategic points around the world, with a map being done of similar temperature of the outer shell at a level of three miles below the surface. This would take a lot of work and time and money, but would enable us to accurately predict earthquakes in the future.

It is not just a matter of one earthquake having a counterbalancing one. It is more like a jigsaw puzzle because the counterbalancing one will then have a counterbalancing one, and so on.

Time, not just location, is also a factor that will be able to be predicted by the temperature monitoring. Hotter temperatures will cause a slower reaction as far as a counterbalancing action.

The variation of temperature between quakes needs to be studied very thoroughly and degrees of differences marked on all the locations of past quakes. This, then, would show the pattern of timing based on temperature differences. This will be further into the future, but is marked on all the locations of past quakes. This, then, would show the pattern of timing based on temperature differences. This will be further into the future, but is the only way for exact predictions to be made. Many lives could be saved this way and much devastation could be avoided because care could be taken to not allow certain areas to be populated.

This same map would then allow volcanic action to be monitored. There is a close link, hence the taking of the temperature of the shell, as given. This releasing of volcanic action will lessen the pressure that causes the quakes. At times it is difficult to tell which pressure release will take place until just shortly before it happens.

It is a matter of timing, you see. If the temperature of the shell is cooler in a certain area than the previous quake or volcano, then the next release will be faster to come than if the last one was at a hot spot in the shell. When it is at a hot spot in the shell, the fluidity of the shell allows for expansion and movement. When in a colder area, the hardening of the shell matter does not allow for this expansion and contraction process, and the release must come rather quickly.

There are other monitoring indications as far as the oceans go, but they are more complex and not as easily reachable as what has been given.

A charting system world wide of venting would show

a pattern of balance. When this pattern of balance is no longer apparent, then will come the time. So be it; it is nature. If the balance is interfered with it will cause major problems in some other manner. A manufactured venting could be accomplished when the patterns are established, but it would not be wise to attempt this as it will upset nature's delicate balance in some other way.

Part III

Personal Cataclysms

Chapter Seven:
Need for Change

Third dimensional life has been a very difficult stage and is getting more intense. The vibration of humanity is in disharmony with the vibrations that Earth is moving into as it goes towards becoming fourth dimensional frequency, which is a higher vibratory rate. The higher the energy on Earth gets with humanity not keeping pace and rising up with it, the more it is pulling the planet apart.

It's a stage that has to be worked through. There's no way to skip it. Some people will require more lifetimes than others to reach the same level of evolvement, but all will have their hundred and forty-four lifetimes for this stage of the path.

Schooling on the Earth plane should be twelve months for each of the twelve grades, instead of nine months for twelve grades that our educational system is based on, to accumulate a hundred and forty-four units of education.

We are shorting ourselves by three units each year, amounting to being deficient by thirty-six units when we graduate from high school.

If a child in school requires eight years to progress through the sixth grade, he will then have only four years left to complete the next six grades. The same is true as far as your spiritual progress is concerned. Perhaps there will be a shifting of more lifetimes to reach the lower six planes and more work required in fewer lifetimes for the upper six planes.

Just as a child in school is held back when he fails a grade, you will be considered a laggard if you do not attain what you set out to do. You will have to repeat the third dimensional Earth plane before progressing to the fourth dimensional Earth plane, which is the true reality for the Earth plane. Earth is the fourth planet, the fourth chakra in this solar system body of the Christ Consciousness. This planet is for struggle and the goal is harmony through the conflict of third dimensional life on the fourth planet. Through this conflict you will achieve the harmony.

Now is the time to reach for and achieve harmony through this conflict and you will move from the third dimension to thc fourth dimension. The transition from third dimension to fourth dimension must be made before any can enter the fifth dimension. People will always be split between the third and fourth dimension of the Earth plane, but everyone must make the move together to the fifth dimension.

The third dimensional experience is a very long and difficult part of the stay on planet Earth. This stage must be understood as a prerequisite to attaining the fourth dimensional Earth reality, the unmanifested stage of Earth's experience.

As planet Earth goes through its changes making the transition from third to fourth dimension, so also will humans. In fact, the planet cannot make the transition until humans do. Actually, they both make the move at the same time. It is a reciprocal process. Progress in one area spurs forward progress in the other.

Humans were formed out of the material of the earth. The very dirt we walk on, the very water we consume and the air we breathe all go to make up the composition of the human form. We are the same as the land itself only a different expression of the Creator's spirit. Each expression of spirit is different but none is better than the other. We humans have been taken further through the creation process to enhance our particular evolution. The lower kingdoms have not had their turn at this yet, but the higher kingdoms have. It is only a matter of time, as we think of it, before the animal kingdom will be uplifted, followed by the botanical and mineral kingdoms. This must take place before the elemental kingdom can begin the next octave, the eighth kingdom.

Everyone needs to be brought into fourth dimensional consciousness not left in third. Some books attempting to do this can be very accurate but from a blunted standpoint. By this I mean that if the author closes everything over, he leaves no open space at the top of his concepts for expansion. If he is working very close with textbook learning he truly needs to bring in the intuition and the stretching of his own mind. Those working this way (using textbook learning) do not realize that they limit evolutionary growth. What they write and set forth to others as absolute concrete facts and knowledge makes the possibility of limitation that much more likely to take place. They must learn that, "Thoughts Are Things."

Humans are at the separation time that has been spoken of erroneously for two thousand years. This separation time was misinterpreted from Jesus' words that He would gather His own to Him; that two would be standing side by side in the field and one would be smote down and the other lifted up.

The time of separation that was being referred to is the present time. Those being lifted up are not being "taken to heaven." They are being raised in their consciousness by the calling of all souls for whom this is a reality. Those being "smote down" are not being knocked to the ground or killed or anything like that. The misinterpretation was very bad when the translating of these words was done. What was meant was that they were not ready, as far as soul accomplishment, to open up to the new consciousness, which is as though being passed from third grade mentality to fourth grade mentality.

It is the difference in third dimensional consciousness and fourth dimensional consciousness that is doing the separating. It is soul dictated. If one is ready, he will be "gathered up," if not, he will be "smote down." It is as simple as that. Those gathered up are directly responsible for the changes that are needed on the planet to raise the planet to fourth dimensional consciousness. When all those who are "gathered up" raise their vibratory fields, they need to help raise the vibratory rate of Earth itself. At that time, then, those who were "smote down" will be able to rise up also. Those being "gathered up" are the leaders; those "smote down" the followers. It is all a matter a soul evolution.

I keep talking about third and fourth dimensional consciousness. To help with your understanding I would like to explain the meaning of some of the terms used in explaining the difference between the third and the fourth dimension.

The meaning of the term consciousness is the progression of the soul. It designates how far the soul has progressed towards building a conscience that will not allow it to commit any act against another that he would not want committed against himself. To achieve the Christ Consciousness, then, would mean a soul of such great spiritual growth that it could commit no wrong against another. A thought such as this would not even enter the mind of one who has achieved the Christ

Consciousness.

The word dimension is applied to the mind of the personality incarnated on the Earth plane, a measurement of the thought level of the new personality.

The word plane is a term used to designate the evolutionary progress of the physical body.

The word planet is to designate exactly which planet the energy is focused on at any particular time. By this I mean the spirit or piece of the Creator that is part of each.

Jesus was of third dimensional consciousness when He entered the Earth plane. Earth is the fourth planet. Therefore, His goal was to rise from third dimensional consciousness to the fourth, which is the goal of all on the Earth plane today.

His physical form was what allowed Him to perform miracles. It was from a planet that was higher than any used in our solar system. The genes impregnated into His mother were from such a high planet that His ability to create high vibrations were the means by which He was able to perform miracles. His ability to create these high vibrations then allowed His mind entrance into all the higher planets where the energy of all divine sparks is focused. This allowed Him insight into humanity and all its problems.

If anyone today had these same genes they too could perform miracles. None on Earth today are capable of what He did, although some people can raise their vibrations to a very high degree and accomplish beginning feats that would be a start to approaching miracles. The time simply isn't right for this sort of injection of genes yet, but will be, perhaps sooner than we thought. Humanity must do more on its own before the next injection of genes can be given. This next injection of genes will raise the vibratory rate of the human race to the level of the form that Jesus had.

Anything that one can do to raise his vibratory rate in any of the four parts of his body will certainly help him on his spiritual path, but none can expect to emulate what Jesus did. This is the reason why none today are able to follow His teachings exactly and take full advantage of His wisdom. The more they raise their vibrations the more they will be able to understand His teachings.

Humanity is in the embryo stage, growing but not fully developed. It is still attached by the umbilical cord to Mother/Father God, and needs to have this nourishment to continue to still grow. Once Earth becomes a sacred planet, then this cord will be cut. You must survive by your internal resources to be qualified to enter the fourth dimension. You must take this step of cutting the cord, ridding yourself of the need for external nourishment, the outer God. You must realize that God is Within.

There is much confusion about the nature of the family

these days. The family as we know it has been with humanity since the animal days. It will always be with physical evolution. It is a third dimensional function of human life – physical life. The family is needed and necessary for those still living and working in the third dimensional world, but for those striving to live in the fourth dimension of the mind, it is a prison. It hampers all thought processes as to further spiritual evolution. Physical ties are no longer important in the fourth dimension. This is a very difficult stage for people and most are not ready to give up the physical ties at this time period. They will simply have to reincarnate and try again in their next life.

All relationships in the third dimension will follow the pattern of evolution – that is, relationships will move from physical growth to emotional, mental and spiritual. It has been said in the past that if a person doesn't relate to another in all four areas, physical, emotional, mental and spiritual, then he does not belong with the other. This is the ideal way for any relationship to be, but that's not to say that life on the third dimensional Earth plane is always going to be ideal. Perhaps life in the fourth dimension will be more ideal in this respect, but almost everyone is still living in the third dimension.

You must learn to relate to those you come into contact with in your daily lives. You must also learn that you cannot repress your feelings, your emotions. You must let them out. To repress your feelings is the worst thing you can do as far as health goes. There must be a release through the creative arts, which is the only way your soul has to express.

The only way human relationships can ever smooth out is for each individual to attain self-esteem before forming a relationship with another. This is imperative and it must be started in childhood. This can be said over and over until you get tired of hearing it. When each one of you has reached synthesization of your higher and lower natures, most relationship problems will be solved automatically. As it is now, each one of you is looking for your own other half, or higher self, if you will, in some other human being. You are constantly disappointed because this other human being is also a "lower self" looking for his "higher self" in someone else. When you realize that what you crave – to find someone that can make you feel whole and complete – can never happen until you find it within yourself, then your progress in human relationships will jump forward. Until this takes place, it is almost futile to hope for improvement in any relationships on the Earth plane, whether it is between individuals or countries. Improvement cannot come until this synthesization is at least understood, even if not accomplished. This is the goal for the fourth dimension, understanding of this fact. The goal of the fifth dimension will be that of accomplishing this fact.

When you finally understand that you will not find what you are looking for in another, and accept that fact, truly accept that fact, you will stop demanding the impossible in your relationships. When you accept the fact that this other one is incomplete within his own make up, you will not have expectations of perfect behavior and correct solutions to all problems.

The marriages and divorces on the Earth plane are outward symbols of the internal struggles of those involved. The internal struggle – that between your half of the soul in manifestation and the half on the inner planes – is the true cause of all unrest in the world including that between husbands and wives. The opposite partner becomes a symbol of the other half of your soul and the duality of each individual is doubled because you are dealing with two sets of duality, that of the husband and the wife. With each being a symbol for the other, the battle with self is played out in real life.

There are many stages of this battle. When you overcome the one last hurdle and become truly mated with yourself, none will be able to shake you loose from your belief in yourself. When you have accomplished the battle within yourself you will then work on the aftermath of the battle and will soon clear up all the stuff around you. You must realize that you went through the battle over and over and over and over again in other lives, sometimes with many different partners, to reach the point you are at.

Some people have not played out all their battles due to limitations imposed on them by themselves, by closing off the emotions and not battling with their partner. If you really battle it out you will both advance much further along the path than otherwise. If you do not battle it out, everything that you should have been battling about will come to the surface later and manifest in different ways. You can see that some people are only acting a role to serve as your teacher.

You need to study and then see how you apply the time you spend with your spouse or even the time away from your spouse. Are you being present in the moment when with him? Could this focus allow you to see things in a different light? No one can say what it is you are to do about your relationships; only you know the answer to this and it will come from within.

Each of us must go through our “forty days and forty nights” of being in the wilderness. Maybe the time is right for your wilderness experience before making the decision as to whether to stay or leave a relationship you are in. How you accomplish this is up to you. The teachings of the past will not help you in this instance. It will only confuse you if you are on the spiritual path. There is no way of turning around and going back. You must determine if you can hold onto your own spiritual beliefs and values and continue growing while in this relationship or not. You must prepare yourself for your eventual crossing as you will do that crossing by yourself. None other can help you with it or with the preparation for it. Your spouse must also prepare himself in any way that is meaningful for him.

Those of you who say that it was love at first sight and

has been love from then on, perhaps earned this as a reward for some achievement in another lifetime. If so, then it was planned that you have smooth sailing for this lifetime. This is a very rare occurrence at this stage of third dimensional life as you can see by looking at the people around you. When we go into the fourth dimension it will not be rare; it will be normal. The struggle that so many are going through now will be rare in the fourth dimension, but this is third dimensional life. This is where you are at the moment and these relationships must be worked out. It must be a struggle and you must put forth effort. The more effort you put forth, the more effort will be instigated in your partner, but one of you must be the catalyst. Then it will become a circular thing and will involve others in the family, and a beautiful relationship will work out. It can be a very beautiful relationship if it's worked on together.

The idealistic view that many of you have of a perfect marriage is childish, wishful thinking. It's not facing reality. Reality is the situation that all are in now and your growth in evolution will come from the struggles in those relationships. Our society had a lot to do with setting these ideals of the perfect marriage with the perfect mate. Our fairy tale movies and our fairy tale books are painting a picture that isn't there and shouldn't be there in the third dimension. That's not what the third dimension is for.

The time is now for an end to outdated emotional reactions such as resentment, self-pity and martyrdom. It is time to be free of that imprisonment. The main thrust of the shift in consciousness has to do with transmuting lower forms of emotions. The emotional body must "clean up" its act or it cannot withstand the pressure of the increase in vibrations and there will be emotional breakdowns in many. It is very important that not only psychologists understand this, but that all others understand this also. This is why everyone should have a creative, artistic release to act as the clearing and cleansing factor in transmuting lower-life forms of emotions. These would be things such as anger, resentment, pride, jealousy, bigotry, prejudice and falseness, and the reactions they cause – fighting, raping, killing, maiming, etc. These need to be transmuted into the higher-life forms of emotions such as compassion, love, joy, hope, integrity, positive reassurance, etc., that will bring forth reactions that will astound humanity.

In the process of clearing and transmuting these emotions and the reaction these emotions cause you to have towards others, you will be progressing your own spiritual standing without even realizing it, and will reap the rewards in many glorious ways. You will find yourself, almost effortlessly, becoming a part of the Creator in a way that none can understand at the present time. There is no feeling like the peace that is experienced when you are finally released from the lower emotionalism. It is time to restructure our whole world because it has been primarily based on emotional reactions.

To review, life in the third dimension is still third dimensional consciousness. You see it on all levels. From countries right on down to individual families there is fighting and arguing taking place, with storms building up. When it's taking place, and all those aspects of nature are manifesting in humans, it affects planet Earth herself. She then takes on the storms, the frustration, the anger, the boiling, churning emotions, and they culminate in an upheaval of some sort. It has to happen to clean the air. It's the same as anger building up in a person. The forces that build up and cause tornadoes or hurricanes have to be released. When Earth releases, there will be a chain reaction, right back through the family. This release will come for all.

Where must it all start? With each and every individual that walks the face of this planet, now and forever more. Think about your past and those associated with you. Remember the harsh words, the violent tempers, the angry tears. Look back at them and remind yourself daily that they will be no more.

There can still be feelings, perhaps even feelings of remorse for past deeds, or the feelings of frustration we have, for instance, because of the fact that many children are still hungry. But even the gentle tears that these thoughts bring will soon be replaced with determination to right those wrongs, those discrepancies that we are able to right.

When all rid themselves of jealousy, anger, greed,

selfishness, and pride, then the natural cataclysms will be over within the planet. These are the esoteric attitudes that must be transmuted. They must first be overcome in the mind, then applied to outer actions. The first step will eliminate internal strife and disruption in the planet, manifesting in peaceful atmospheric conditions. The second step will eliminate violence and angry words and killing, raping and maiming of your brothers and sisters. This, then, will manifest itself as a ceasing of war and crime. There can be no more wars, as this violence will no longer be possible without total destruction, not only of humanity, but the planet herself. Only after these two steps are taken can there be hope of peace. Think about an existence where all are at peace with their neighbors and, more importantly, with themselves.

Those ready will be lifted into the higher planes at night and trained to enter the fourth dimension. It's going to be much harder for those who are lifted up and trained to start the new culture in the fourth dimension, than it will be for those who are not. Those who are not will make the adjustments necessary for life in the fourth dimension on the inner planes after they leave this incarnation, if they can't accept what is taking place now on the Earth plane.

Picture humanity as one small guppy being born at a time. Get an overall view of humanity after birth, growing through maturity and reaching the "golden age." It really is a long way off, but with each birth (each person seeing the light of truth), it comes closer.

Compare each year of that newborn babe's life with one hundred years of Earth time. Even the most advanced (in evolution) are not much past the toddler stage in comparison to a baby's growth. A few are at kindergarten age. This is just a figure for comparison with the distant goal. They will enter the fourth dimension one at a time, (each birth, each awakening) and work and grow slowly towards the "golden age" of mankind. Consider, then, how long it will take to reach the "golden age." Remember, the "golden age" will last one thousand years; then the death of our star will be imminent.

Many of the ideas that are presented by these Chrysalis Teachings are for those of you who already have made the breakthrough into the fourth dimensional mind level. You can be compared to the first class of students in a new school and this term is over. Now you must move up to the next grade and allow a new first grade to enter the awareness arena. Meanwhile you'll do all you can to implement what you learned in that arena to help those just starting down this most difficult path. It will be easier for them because you have blazed the trail for them, but before the breakthrough is made, they may not understand many of the concepts presented by these teachings. They will still be trying to take responsibility for everyone and everything except themselves. After they make the transition into the fourth dimensional mind level, they will realize that they are the only one they are responsible for.

They will also need to understand that when one begins

working on himself, competitiveness is one of the attitudes that will be removed. As always with people, the pendulum will swing too far the opposite direction and they will find themselves with a total lack of competitiveness, whereas before they had too much competitiveness. It will take several years before the pendulum swings back to a balanced center point.

A certain amount of competitiveness must be present to enable you to take responsibility for yourself. The problems usually come when you try to take responsibility for those around you. This is what causes over-competitiveness. When this is eliminated and you realize truly that you must be responsible for yourself and only yourself, then will the lack of ambition to do anything be corrected. You will once again take your place as productive members of society. Until then you will have to battle with the terrible lethargy that overtakes you, the feeling in your mind of not being able to concentrate and not being interested in anything.

It has been my experience that those making the transition from the third to the fourth dimension enter a state of "limbo," so to speak. I have stood by and watched as many of them sort of drifted off the path and ended up doing nothing for a period of time. They seemed quite lost, as if they didn't know what the next step was for them so they just decided to take no steps at all. A lot of them moved around, going from one place to another searching for something that felt right. There are still many people in this stage. These are people with brilliant minds, for the most part. All their friends can do is watch and wait for them to find their way back.

This lack of ambition in some of you on the spiritual path is a natural outcome of working on yourself to eliminate old attitudes, actions and reactions. Therefore, lethargy is a part of the spiritual path, but it is the part that is present just before the breakthrough comes that will lead to total transformation into the fourth dimensional mind level. Once you make the breakthrough there will be no stopping you from achieving major changes in your own life. This will also affect, positively, the lives of all those you are in contact with.

There are many of you who have not reached this stage of lethargy yet. You are still working on projects that you took on in the third dimension. You have to finish these projects before you will enter this state of limbo. This state lasts for different lengths of time for everyone. Therefore, there is no way to predict how long each one of you will be in it before you totally make the transformation into the fourth dimension. This will depend a lot on you and whether you fight the state of limbo or accept it.

It is quite overwhelming when one finally reaches the point of being able to see a much broader picture of Earth and her inhabitants. It seems impossible to try and make the changes that need to be made. You need to have the period of time that the stage of limbo brings, to sort it all out and make some choices about what you can accomplish and how you should go about accomplishing it. When you do find your way back you will be a very productive member of society, working to help others make the transition and working at the same time to improve conditions on the planet.

But first, you have to go through a battle with your animal-based brain not wanting to bow out and give up authority to your, by this time, well-entrenched soul. There are two different steps to the process. The first step, the integration process, must come before the second step which is the actual move of your soul into the physical organ chosen before entry for it to be embedded in.

To integrate two things is to get them working together or vibrating at the exact same level. This vibratory rate must be exact in order for the tiny electrical spark to be able to penetrate physical matter and lodge in the physical organ.

It is the same process as giving birth in humans. The contractions must begin vibrating in exact harmony with the pushing of the fetus. This will allow the birth to take place into the thick atmosphere of the planet. That process takes a period of integration of allowing the contractions to reach the correct rate. It is the same with the vibrations of the body and soul. They must be in absolute perfect harmony for the "birth" into the body of the electrical spark to take place. This is truly a period of rebirth and is cause for great celebration when this takes place.

Many of you are going through the stage of confusion that all go through before breaking out into the true, beautiful, shining light of the fourth dimension. Until all have made the transition, the world and everyone in it is going to be going through a state of chaos. When each one has made the final transition into the fourth dimension, then will come the needed insight into how to structure our society so all can function as individuals. In other words, individual rights will be the basis of all forms of government in the fourth dimension.

This transformation, of course, affects all relationships, whether between just two people or between countries. For this reason peace will not, cannot reign on Earth until fourth dimensional thinking is reached by all. It must start in the field of education, for adults as well as for children. This is why it is so important to get these teachings out. The material in these teachings will benefit those of you seeking to become better people. It will allow you to ponder on the fact that we are not alone in this universe; that there is structure to our path and definite goals ahead to continually strive for. This is all documented for others to study; the whole story is laid out in the Chrysalis Teachings as a map for everyone to follow. These teachings will help psychologists to know at what point on the spiritual path people are standing when they come to them with problems. All mental illnesses and mental aberrations are made up of the standing of a person on his spiritual path.

The fourth dimension is a clearheaded space where there

are feelings but not raw emotions. There is a big difference. You will still have periods of slipping back into the third dimension when drawn into other people's space. These must only last for a short period of time. You must raise yourself right back up again into fourth dimensional thinking by whatever means you can. You will find that you no longer wish to be with those of third dimensional thinking, but you will have to remain in the third dimension while working with others. While being with them you will need to always keep one foot in the fourth dimension and be able to draw yourself right back up to it when you leave their presence. You actually can keep part of the fourth dimension with you as an overview of the situation while performing in the third. This is the proper way for the transitional period to be handled.

By transitional period, I mean the period of time when all are mixed between the third and fourth dimension. This mixture will be with you the rest of this life and perhaps all of the next and maybe several more after that. It will be with you until enough people make the transition to the fourth dimension that there will be no problems in dealing with them in everyday life.

For now, everyone operating in the fourth dimension must live a split life in the everyday world. You can be wholly into the fourth dimension when not having to deal with the outside world, say, for instance, on the weekends or evenings, but this would tend to isolate you.

You must be able, at all times, to live a split life. You

must be able to understand the thought processes of the third dimension and have compassion for those still expressing through this behavior.

At the same time, you must always come at problems from your fourth dimensional thinking and then step the solution to problems down to the third dimensional world. This is the best training ground for acquiring the use of wisdom. You can have much knowledge of all subjects, but if you cannot apply the knowledge in a wise way to third dimensional living then it is useless and will cause you a life of misery.

One particular area that most people need to work on is that of letting their light shine for others. You all need to become more outgoing and externalize your spirit. It has been kept under cover for too long. This light is beautiful and it should be let out in front of more and more people by being with and relating to others. You all need to have your own inner strength of knowing that you are your own source of power and inspiration. You need to trust the very core of your being and then follow your intuition. You all do a lot of living in your mind but you need to do this living on the physical plane also. Follow your wants and let them lead you to fulfillment as a spiritual being, no matter where this may lead you.

Know that the New Age of Peace is being born and you are part of the birth pangs. You must also be responsible for the tender nursing of this wee fragile bud of peace and double your efforts in the years to come to help it reach the full bloom of the Rose of Peace. Let no more fear of ostracism be felt by anyone in the new consciousness because you are the rightful rebuilders of the Earth.

You who are passengers on this beautiful ship in space are those who must fulfill the destiny of the planet herself. Her destiny is to be finished with violent storms and responses to inner change, the same way we are to be finished with violent emotional storms and responses. Henceforth there should be gentle cleansing rains, not only for the planet but also for us. These gentle cleansings can only come after the violent whirlwinds and deep earthquakes that have shaken us through all time thus far are over. The planet is ready for change. Are you?

Chapter Eight:
The Fences of our Mind

The true fourth dimension is like a "halo" around the planet for all to aspire to. This would be to reach the purity and cleanliness, mentally, of fourth dimensional thinking processes. People must synthesize their lower nature with their higher divine self before they can enter the fourth dimension because, as discussed in the previous chapter, none of the lower nature traits and attitudes will be allowed in this fourth dimension.

The preliminary stages are past, those of awakening to the reality of our precarious situation. Next comes the implementing of the esoteric lessons learned and bringing forth this learning into your exoteric life patterns. This will mean breaking old patterns and replacing them with the new – a tremendously difficult thing to accomplish. But know that you will always have help and support from those in the higher realms in your efforts to implement the truths that you have gained. You must realize that you are facing a very difficult transition changing from third dimensional consciousness to fourth. It takes much perseverance, and obviously a soul commitment, to stand firm against the tremendous pull that society has to "suck" you into the established ways of living your life. This pull is very hard to swim uphill against and most can't accomplish it due, I suppose, for fear of being an outcast from the "tribe." Isn't this a throwback to the days of early man? Are things really much different now than they were then? Perhaps they are on a more civilized basis, but the result of societal patterns is still much the same. They tie up the minds of many that could be free to wonder and grow and evolve more rapidly. It, the establishment, uses fear of ostracism to hold us inside the fences of our mind. The present time can be considered a "grace period" for all to eliminate these fences that not only separate your creative, intelligent and spiritual mind levels from your physical life, but also hold you separate from the rest of humanity. These fences must come down in all areas. Now is the time.

Ways of bringing this tearing down of fences into being are many. I have already mentioned the benefits of a creative, artistic release. Other ways would be to read a book that is "out of the norm" for you to read. Set a goal of perhaps one "odd" book every two months or so. Another way would be to buy an article of clothing that is not your normal way of dressing. Perhaps you will feel self-conscious and uncomfortable in it, but force yourself to wear it. This will be breaking fences.

Cook and eat one different dish a month. Make a party out of it. Perhaps you won't like it, but picture the fence crumbling and make yourself eat it. Go barefoot out to the mailbox once a week if you always wear shoes. Try a new way of fixing your hair, or a new brand of toothpaste or vitamins.

The worst thing you can do for the health of your body is to become set in your ways and get stuck in the same old patterns. You need to become interested and involved in cultures other than your own. This will broaden your mind faster than anything else I can think of. You will soon discover that the way you deem best for all to aspire to live their lives is perhaps considered an atrocious way to those from other countries. This will have the effect of loosening the hold that the animal-based brain ego has on everyone. This applies to all countries, not just ours, although our standard of living is much more pleasant than most others. This should make us more obligated to want to share ideas with those less fortunate.

Set aside one evening a week, or each payday, for an evening of enlightenment in some form or other on a new subject. Attend a lecture about growing rice in rice paddies in China, for instance. Do it by means of TV or videos if necessary, but do it! Broaden your horizons in any way possible! Use your mind!

As Earth makes her move through cataclysmic activity, so also does the human. This cataclysmic activity in the human takes place in the mind/brain. To help you understand the process that must be gone through, I would like to explain the functions of the brain and show how the move is made from third dimensional thinking to fourth.

Until now we have only been aware of the difference in

the functions between the right and left side of the brain. It is now time to become aware of the different functions of the lower and the higher mind in conjunction with the right and left sides of the brain. Each side has an upper and lower division.

The brain has four sections with three levels each. Each of the three levels is then subdivided into twelve divisions. The first section, belonging to physical body functions, is the lower left and is made up of levels, 1-2-3. The second section, the emotional body section, is the lower right brain and consists of levels 4-5-6. Levels 7-8-9 make up the mental body section, the upper left section of the brain. The fourth section, belonging to spiritual body functions, is made up of levels 10-11-12 in the upper right part of the brain. Therefore, I'm talking about twelve levels of the human mind. Each level then has twelve subdivisions.

Most of humanity is functioning on the middle to upper subdivisions of the third level, in the lower left section belonging to physical body functions. You have heard it said many times, I'm sure, that you're only using ten or fifteen percent of your brain – maybe twenty or even twenty-five percent if ready to step into the fourth dimension. The rest of your brain power is in these other three sections of the brain.

The second section, the lower right 4-5-6 is where all should be operating by now, but only a small percentage are. As just explained, most of humanity is on the third level in the first section, the lower left section – that of the physical body – and must be able to at least reach the

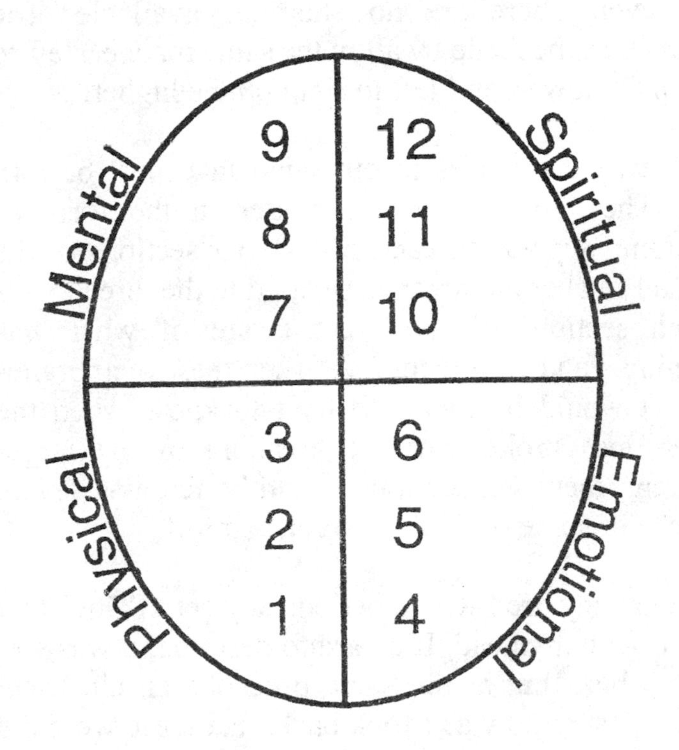

The Brain

The four sections of the brain showing the twelve dimensions of the mind. This is the structure of the brain and the path of progress should be from lower left brain to lower right brain, then to upper left and then upper right. The left deals with exoteric functions (daily living), the right with exoteric functions (the creative process). – Illustration by Neil Hickox lowest level of the second section, that of the fourth mind level. There are no shortcuts available. The progress must be made by all in the same manner, left to right on the lower, and left to right on the higher.

There really are fences in our mind that must be torn down. They are in ethereal matter in the form of membranes separating each of the four sections of the brain and smaller membranes separating the three levels in each section. If you were aware of what was happening, you could recognize when these membranes broke. It would be more difficult to know when the smaller ones broke, but the breaking of the larger membranes between sections would be felt as a severe headache or some other form of discomfort.

All through my life I would periodically get a "zing" that went through my head. It's hard to describe; it was sort of like a buzz but at the same time like an electrical shock. I'm sure now as I look back, that these were the breaking of smaller membranes.

I also distinctly remember when I broke over into the second section of the brain – from the third dimension to the fourth. I had an awful, horrible headache, worse than any I'd ever had in my life. I lay down on the bed with a wet cloth on my forehead. Soon after that I had the most beautiful vision, in color, of what the city of Grand Junction would look like in the future. The Colorado River had either moved further north, or had been diverted, to flow further north. Either way, it ran right through the main part of the city. All the light poles glowed with brilliant colors and reflected in the water. I can't remember too many other details now, but it was beautiful. I have never had such a vivid, colorful picture of anything since then. It took several hours for the headache to let up. I thought to myself if that's what it took to have a beautiful vision like that then I didn't want any more; the pain had been that intense. That thought is most likely why it has never happened again. I stopped it myself.

There are many names given these sections of the brain by different people, and other names are used by psychologists. Because I don't want there to be any confusion as to what I'm are referring to, everything will be based on material given in the Chrysalis Teachings.

The actual scientific meaning of the word enlightenment means the breaking of the membrane between each section of the brain. In the previous chapter I talked of the stretching and reaching that we all should be doing. When one is totally enlightened, each section of the brain is open to this stretching with new thought. Any section that has not had the membrane broken will not be open to this stretching. It is held in tight constriction and cannot allow any new thoughts to enter.

To further your understanding, I'd like you to picture a tunnel that is well lit at the entrance. It has three branches off from it a short distance from the entrance. These branches are sealed over with a canvas that prevents any of the light at the entrance from penetrating into these other tunnels. Until there is light inside them they cannot be enlarged and no good can be gained from having these other tunnels. Once the canvas that seals the opening is removed, men and machines can enter the tunnel and enlarge it and cause it to be able, eventually, to connect with the next tunnel until the whole mine is operating at full capacity and the goods found in each tunnel are being reaped for harvest.

These divisions of the brain are noted esoterically by scientists, but not exoterically. There has not been much dissection of brains done on humans. Most has been done on animals and this is one place where the animals cannot serve the scientists because they do not have these membranes. They could be found and noted if looked for in a human brain. In other words, if a research project to this effect were begun, the scientists, by using "black light," would be able to identify these minuscule membranes. With ordinary light and eye sight they would not. An instrument combining "black light" and magnification would show them.

On occasion someone will hit their head and suddenly have unusually strong mental powers, or psychic abilities. The reason for this is that the blow to the head causes the breakage of the membrane between the first and second sections of the brain. Usually, when one has an accident that causes the splitting of the membrane, it gives instant access into the psychic arena of the brain: the astral plane. If you will remember the drawing of the memory pole in chapter two, this is the area between the third and fourth dimensions. This then allows the breakthrough from third dimensional thought into fourth dimensional thought to take place.

There is an exact pathway of these levels of thought that

will show the road map for others to follow. It is not a wise idea to try to cause this splitting of the membrane to take place on purpose. Much damage could come from any attempts to do this, both physically and mentally. The only way you could do this would be by a blow to the head that could cause severe damage in other ways. It will open naturally when it is ready. It cannot be achieved by surgery, at least not physical surgery. It is possible that a psychic surgeon could remove the membrane.

There is one thing that will help with the breaking of these membranes. The vibrations set up by the musical tones of a piano or other acoustic instruments, especially stringed instruments, will serve to separate the membranes between the layers in the brain. (In my case, I have been playing the piano since the age of four which would account for the breaking of these membranes.)

This membrane is of ethereal matter as all membranes in the brain are. It is not like the physical membrane found in a virgin female. That is made up of physical matter. These membranes are of ethereal matter because the whole realm of the brain is the ethereal realm. The physical components of the brain are merely holding agents and sectioning agents.

You could compare these membranes or "fences" in your mind to the walls of a school where the school is divided into classrooms. The outer walls of the school are the holding agents; the inner walls (membranes) hold the classes (different levels of education) that are made up of physical bodies.

To compare the brain to the school, the brain is the holding agent for ethereal thought; the membranes doing the sectioning hold the classes (different levels of brain power) that are made up of ethereal bodies.

In the school the different classes would still be intact if the walls were not there dividing the physical reality by sectioning the space off. In the brain the classes or levels of brain power are there in thought, even if you cannot see the membranes – exactly the same as in the school.

The progress of evolution takes a major jump at different stages in history. It's constantly climbing but there are always little platforms which are what these membranes are. As given, each section of the brain has three levels and each of these three levels has twelve divisions. Levels one and two are closed off (due to the dropping off of these two dimensions of thought when we entered the third dimension). You will progress through twelve divisional breakdowns in level three. When you come to the twelfth division in the third level of the brain, you're going to a different section of the mind, the second section; therefore, it is a much bigger jump. This is the point that humanity is at right now, making that huge leap forward from the first section to the second section of the brain.

This is the transition stage that adults from the approximate age of thirty on up are going through. They are the actual transitional humans. This is not to say that those younger than thirty are true fourth dimension. It's just that their physical forms are more easily adaptable to fourth dimensional vibrations due to the atomic make up used in forming the flesh.

The state of humanity is in a changeover period as big as the change that took place after Atlantis. At that time the chakras were rearranged and aligned and the atomic structure was changed. The same thing is taking place at this very moment in humanity. The most difficult time to be aware of a transition is when you are in the middle of it. If you were looking back at this period from the future you would be able to see that it is another quantum leap, the same as the restructuring that took place after Atlantis. As that restructuring was preceded by cataclysms, so also is this restructuring – both planetary and personal. We all survived that period; we will survive this period also.

The more you understand the process that is gone through, the easier the transition from third dimension to fourth will be. Let me explain further about the levels of thought.

Perhaps some of you have had an opportunity to see someone bend a spoon with mental thought. Bending spoons or making clocks work is a higher level of third dimensional thought. It is approximately the tenth division of the twelve divisions of the third level. The entry into this level gives a "shot" of power to the brain. It is not sustainable, or rather should not be sustained because of the damage to the physical body and the inability to move past it, to "graduate" into fourth dimensional thought processes. Many people get caught and locked here on this level demonstrating their feat of bending spoons or whatever. This will eventually damage the body and cause disease.

You see, when you get into the higher levels of third dimensional thought it is imperative to keep moving because of the intensity of the vibration that could possibly cause damage to the physical substance. Even though all third dimensional thought is harmonious with the electrical system of human bodies because it is still accessing the proper electrical synapses, staying in the higher intensity could cause trouble. The newer forms entering the Earth plane now will not have much difficulty with any of this.

People go through all twelve of the subdivisions of the third dimensional thought level and then must go through the twelve subdivisions on the fourth. When you start working your way up the divisions of fourth dimensional thought, there are new synapses being accessed and the vibratory rate of the third subdivision of this electrical current is one that is especially hard on the physical body. It is a vibration that is doing its job by fracturing through the wall or fences that third dimensional thought has you locked into. It must do this or you will not be able to make the breakthrough into the fourth division of fourth dimensional thought.

Most people are capable of reaching the third division of fourth dimensional thought processes for short periods of time. It is the ability to sustain this thought that is difficult, if not impossible. The third division of fourth dimensional thought takes a kind of energy that the human body is not capable of sustaining for long periods of time. It is the transitional period, this third division of fourth dimensional thought, that is breaking through the fences of locked-in third dimensional thoughts. It is very difficult. One must not overstay his time period in this thought. He must periodically leave it and reenter third dimensional thought or it will cause a breakdown in all thought processes. This could happen if the transition takes place too fast with no relief periods from the intensity of the vibration. Therefore, a long period of repair may be necessary.

The fourth division of fourth dimensional thought is a much more harmonious vibration and does not cause the stress on the physical body that the third division of fourth dimensional thought does. Actually, the brain waves past the third division of fourth dimensional thought are very easy to stay in for longer periods of time. You see, once the fourth division of fourth dimensional thought is reached, the vibration "locks" into the energy waves of the universe and you are able to "float" along on this new current.

Even when one is not accessing his inner being, one should still be in the fourth division of fourth dimensional thought for most of the time. The only time they won't be is when they don't take care of themselves or when they dwell on things that upset them. Then they will slip back into third dimensional thought and have to get back on track again.

For those who would like to be able to reach and stay in fourth dimensional thought, it requires a concentrated effort of not thinking negative thoughts. This must also be combined with reaching, at all times, into the higher realms of thought – in other words, trying to make contact with the higher planes consciously. Doing one (not thinking negatively) without the other (trying to reach higher planes) will not accomplish fourth dimensional thinking. You could have been in contact with the higher realms for quite a few number of years and still not have reached fourth dimensional thought processes. When you combine this contact with a conscious effort to think no negative thoughts, you would then be able to reach fourth dimensional thought and hold it there. You will be able to choose to stay in fourth dimensional thought except when dealing with those in third dimensional thought. Even then you will always have the ability to immediately go back to fourth dimensional thought simply by thinking up and thinking positive.

Remember, it is not the content of the thought alone that will produce fourth dimensional thinking; it is combined with efforts of stretching into the upper mind levels. Also, fourth dimensional thought processes are not concerned with group action. It is the fourth plane of separation.

Fourth dimensional thought is not the alpha state and most people do their connecting with their inner being from the alpha state. The alpha state is an altered state of brain waves. By altered state, I mean the electrical current has been manipulated. It has been partially shut down so that the brain is in a hypnotized state.

I don't go into this altered state very often when connecting. From the beginning I did conscious connecting and have only gone into altered states on rare occasions. I broke through the communication barrier before my soul had taken up residence in my body, but nevertheless I did not use the altered state. I used fourth dimensional thought processes for connecting. I always knew I was doing it differently than other people, but never fully understood it until recently.

Altered state connection will hold you on third dimensional consciousness thinking. There is nothing wrong with it, but it does limit your potential for growing into full use of your brain power, at least full use that you have evolved to at this time. Fourth dimensional brain power is by no means the end of growth. There is much to come after that.

There are visualization exercises that will help achieve fourth dimensional thought, but not until after your soul has taken up residence in the physical organ. This will be the blending of ethereal and physical matter. This is a tricky, delicate time for everyone. The spleen acts as a filter or cleanser for the pancreas after the soul enters its physical organ. It does not matter which organ your soul takes up residence in because the spleen is the filtering and cleansing agent. It automatically bonds with the organ being used for the home of the divine spark in physical residence. Nutrition at this time is very important. The spleen is an organ that is thought to be not needed in the human body, but it is a very important organ. The cleaner and clearer the spleen, the purer the material will be. This is a function of the physical body concerning exercise and, as stated, nutrition.

The visualization exercises that will help achieve fourth dimensional thoughts are those that allow your brain to "wander" but not go into an altered state. By wander, I mean give it free rein to scatter its thoughts for a period of five minutes. At that time, then, you are to picture a very high mountain peak. You are to start at the bottom of this mountain with a philosophical question in your mind. Write it down before your start. Then you are to mentally climb that mountain carrying this thought or question in your backpack. You visualize yourself making periodic stops to rest on the way up. On these rest periods you take off your back pack, open it and take out the paper with the question on it and add whatever thought comes to you at that time, if any. Some days you may not receive any. In fact, you may make it all the way up the mountain for the time period of fifteen minutes, and not receive anything new. Other days you may receive a different perspective on the question at each rest period.

You will be confused when the meditation is over, but you must remember, the goal of this meditation is not to achieve the "correct" answer to the question. It does not matter what answer you come up with, if any. The goal is to allow your mind to stretch to its highest level while keeping your focus on the Earth plane and not going into an altered state. This is very important. Do not allow yourself to go into an "altered state mode" or you will not ever reach fourth dimensional thought.

This exercise will open the doors into expanded thought and will eventually help open the door to communication with your inner being, but this must not happen first. It must be the result of reaching fourth dimensional thinking. I did it backwards. I opened the communication door first and have had a much harder time entering the true fourth dimension than those of you will who can achieve fourth dimensional thought processes before opening the doors of communication. This is not to say that your communication with your inner being will be any better or wiser if you do it this way, but it will be much easier on your emotional body if you can open fourth dimensional thought processes first.

Once you reach the ability to handle these finer brain wave patterns, it really is very simple to hold yourself in fourth dimensional thought. You can stay there for long periods of time just by being determined to do so. If you would just make a statement to this effect before going to bed and immediately upon waking in the morning, your brain will take care of the rest.

To review, fourth dimensional thought is not an altered state, but rather an evolved state of the brain. It uses a different level of energy; a different level of electrical current. It is a finer, higher-pitched vibration that is sent through the brain. It is activated by the soul, which must have taken up residence in the physical body by this time. The soul takes up residence in the physical organ that relates to the chakra that the soul has been in until the joining of the soul and physical body takes place. It actually has nothing to do with the connecting most people do, which is a function of an altered state.

The cellular connecting that is being done now (to bring this material through) is not coming from an altered state, but directly from my brain in a fourth dimensional mode of thinking. It is coming from power being sent through my nervous system from my solar plexus to my brain – the solar plexus being the chakra where my soul has been embedded. The physical organ related to this chakra is the pancreas.

You must all simply have more faith and trust in the thought processes of your own brain power. This is imperative in order for you to accomplish all that you want to accomplish. You must trust yourself. Once you are in the fourth dimension you will have a fantastic backup system to your brain, an open-ended ability of accessing the best in thought processes from all of humanity for all eternity. You must use it. It will be available to you at all times. You will need to just think of the problem that you want to have an answer to, or a way to handle the stress that certain thoughts cause. You will then be given exactly what you need to have from this fantastic storehouse available to you.

A test for determining fourth dimensional thought has not yet been developed. At the present time there is no way you can tell if others are in the third or the fourth dimension, other than your own intuition about them.

You are the only one who can tell if you have entered the fourth dimension and if there is any doubt in your mind, then you are not there yet. When you are there, you will know it. Your emotional body will feel enormous amounts of love well up in you at times. Compassion will fill your whole body. You will feel speechless at the time of this great emotional fulfilling. You will not be torn with wracking sobs or be depressed or forlorn or sad. You will be joyful through the emotional welling-up that will be as a communion service for you. It will be spiritual fulfillment to have these emotional experiences, whether they be of love or compassion. They will never be anything else from this point on. So if there is any other emotion that wells up, you will know that you have slipped back to the third dimension. This is when you must do whatever you can to pull yourself back up to the fourth dimension again.

You must realize that to make the transition from the third dimension to the fourth, you will be going through cataclysms and upheavals. You will need to have help when you come out on the other side of the uphill climb. You need to know that you will have accomplished a major breakthrough. You will then need to have guidance and direction as to what to expect and what to do with yourself.

The focus that these teachings you are reading now, the Chrysalis Teachings, have is that of being a pathfinder for the aftermath of the cataclysms you are, or will be, going through. These teachings will tell you what to expect when you have made the transition and how to apply your new gained strength and clear-sighted vision to your everyday life. This is very important; this is the Spirit of Joy.

Chapter Nine:
New Ways of Thinking

Many of you are probably confused as to just what the fourth dimension is. The fourth dimension is a dimension of thought. As long as you are in a physical body the only way it can be experienced is in your mind, your thoughts.

Some people picture an actual physical location for the fourth dimension. This is not so. The fourth dimension is of ethereal matter and cannot manifest in physical matter. The only way it can be experienced on Earth is in your thoughts. You cannot tell the difference by looking or hearing or smelling as to whether someone is in third or fourth dimensional fields, although perhaps they could tell by their overall feelings.

The third dimension of planet Earth seems to be heavy, cloudy, saturated with moisture, bogged down in mud and always on an uphill climb.

The fourth dimension, in comparison, seems to be clear,

sunny, bright, light, airy and on ground with a slight downhill slope. The fourth dimension is a feeling that is upheld throughout your existence. By feeling, I mean not only an emotional feeling, but also a physical feeling and a mental feeling. It involves all three; body, mind and spirit.

The difference in physical feelings would be as follows: as far as the fourth dimension, in your physical body there will be a feeling of lightness, of just plain feeling good. At its height it will feel like a thrill in the body similar to that of cellular orgasm.

In comparison, third dimensional physical feeling could be described as a feeling of fear in the pit of your stomach, or of being out of sorts; feeling sick in the stomach and tired with aching joints. Sleepiness is also another aspect of third dimensional physical feeling.

The difference in emotional feelings would be as follows: third dimensional emotions are those of the lower nature such as low self worth, jealousy, one upmanship, hate, fear, negativity of all sorts, hopelessness and depression. It is a feeling of constantly sitting on top of a volcano that the slightest jar to the ego will cause to erupt. It is a quick stab in the pit of your stomach when anger strikes you over the misbehavior, whether real or imagined, by someone close to you. In fact, much of third dimensional emotionalism is imagined. It is a play-out in your mind of things you believe other people would like to see happen to you.

These play-outs take place because of guilt in your mind over something you did to these other people in the past. It could also be thoughts of what you would like to see happen to them in the future.

On the other hand, fourth dimensional emotionalism is that of compassion and wanting to help other people who are going through emotional upsets. When you are truly in fourth dimensional thought processes you will have no anger or animosity against anyone. You will no longer "want to get even" with another for imagined wrongs, or even actual wrongdoing, that this other one perpetrated against you in the past. All this will no longer be important. What will be important in the thought processes is how to stay constantly connected to your inner being.

By this, I mean to know exactly what it is you want to achieve to further your own spiritual growth. At the same time, there is a desire to apply what you have learned on your spiritual journey to helping others. You want to pull them up to a higher level of viewing not only their own life, but also the lives of everyone else.

Most wars are being fought by third dimensional countries and some of these are at the lowest level of third dimensional thinking. Countries that have progressed to the point where approximately one third of its citizens are in fourth dimensional thinking, such as our country, need to give these other countries what help they can. More important than this, we need to help raise the consciousness level of the other two-thirds of our own country. If the other two-thirds were raised to fourth dimensional consciousness, the whole country would be in agreement as to helping lower-evolved countries.

The difference in mental feelings would be as follows: mental processes of those in the third dimension are slow and belabored when trying to follow a theme through to a logical conclusion. For instance, thought processes that apply to personal problems are very narrow and limited in the third dimension. It is similar to sitting in the branches of a tree and not being able to see its trunk or the top of the tree. All you can see are the immediate branches closest to hand and maybe a small patch of sunlight coming in between the leaves higher up in the tree.

When you are in fourth dimensional thought processes there is a clear and almost instantaneous path from conception of an idea to a logical conclusion. This applies to all intellectual study. Referring to our analogy of the tree, when in the fourth dimension you see not only the tree in its entirety, but the field the tree is in and all the sky above it.

Another example of third dimensional thinking would be someone who focuses on the pain of, perhaps, a swollen joint or an upset stomach or any other physical ailment. By focusing all their attention on it, it grows and becomes worse. The joint will actually swell and become even bigger because of the attention. You see, the cells in the joint are quite aware of the focus on them and they respond to the thought that there is something wrong.

If in fourth dimensional thought, you would simply note, for a brief second, that a certain joint needs to be activated and used more. You would then put it out of your mind after programming your subconscious mind to exercise the joint. It would not be thought of again.

In the case of third dimensional thought, all sorts of garbage is dug up as to why it hurts and what you did to deserve this. You wonder why it didn't happen to someone else, someone who is really a bad person and deserves something like this more than you do.

Mental thought in the fourth dimension will focus on better ways to do things. This is where inventions come from. Fourth dimensional thought always sees a bigger picture while third dimensional thought sees a small, narrow view with everything focused on the self. When in the fourth dimension you know your mind will take care of you and you will be free to focus on matters outside your own world. Your thoughts are unlimited and therefore your world is unlimited. It is filled with possibility at all times.

Quite often, when in the fourth dimension, you will achieve third dimensional goals in your mind and no longer need to physically achieve these goals. This allows you to progress faster and higher on the levels of fourth dimensional thought.

The difference between third and fourth dimensional thought could be compared to the difference in vibratory rate between octaves on a piano. For instance, if normal third dimensional thought brain waves vibrate in you at the rate of middle C, then fourth dimensional thought would vibrate at the rate of the C above middle C. This would be a higher, faster vibration that allows for direct translation of higher thought blocks. There would then be no need of going into an altered state to access and interpret them. The interpretation is present at the time of receiving them. It is a function of an integrated right and left brain that allows the highest brain levels to be reached.

People have the answers to everything locked inside the levels of their brain, but to activate these levels takes the cooperation of all the cells in the body. It is a reciprocal system. Any change that a person wishes to instigate must start with a thought that comes into the brain when connecting with their inner being or God Within. Once this thought enters the brain it can go no further until the lower levels, which form the base of the intended transformation, are laid and refined and built up to then again meet with the thought that formed in the mind.

For example, say a person has the thought that they wish to lose weight. This thought was contacted during connection with their inner being. It could have come about as a response to meditation on a loss of self-esteem, for instance. Once this thought enters the active or conscious level of the brain, it must be transmitted to all the cells of the body. This can be done by visualization.

Visualization is the quickest way to transmit the thought to other cells. It is a “language” that all cells will understand because it is a picture. It is not couched in words that sometimes are too complicated for certain uneducated cells of other organs that may, perhaps, not be doing their part in holding metabolism levels where they should be. In other words, they cannot understand the language used by the brain when writing journals or reciting mantras through spoken language.

But a picture is easily understood. When this picture is given individually to each organ in the body and viewed as to its place in the smooth-running operation of the body, then the education process for that organ has been achieved and one can move on to the other organs and then to the circulating system and the nutritional conductors and the elimination procedures, etc.

To do all this to the point where one can literally change the whole composition of his physical body requires education in the brain of the make up of the physical workings of the body. Therefore, doctors would be the ones most able to accomplish this process for themselves because of their knowledge of the functioning of the human body.

But there is no reason why ordinary human beings can’t

do this, especially if health classes detailing the make up and functioning of all organs and systems of the body were incorporated into early childhood education and, at the same time, visualization concepts were taught. These children would then grow up with the ability to keep their physical forms in a youthful condition until they reached old age. These physical forms would not cause the death of the body. This would be brought about by the mind. In other words, the physical body could remain in use as long as the mind continued to need it. Of course, by mind, I'm referring to the spirit which will dictate the length of the incarnation to the body.

The mind, along with the visual and audio equipment and olfactory senses, is truly a powerhouse. But remember, the mind or spirit cannot achieve the use of this powerhouse without the built-in workings or factory of the physical form – the physical eyes and ears and the smelling ability. Without these, the spirit could not refine visualization to the point where it could transmute physical matter. This is why the transmutation of physical matter cannot be done by the methods being used today.

People are trying to heal and change things with the use of meditation, contacting the unmanifested parts of themselves. It cannot work that way. The mind must have the use of, and cooperation from, the physical body to ever heal the physical body. The spirit cannot do it by itself. If it could there would be many instantaneous healings because of all the prayer that is done.

I'm not saying that prayer is not good. It does help, but not in the way that people think it will. It helps by putting you in touch with your inner being which will then try to send the message to the physical brain as to what is needed to heal. But the healing cannot be done by prayer alone. There is no way this can be proven and there are many people who will never be convinced of these facts, so I'll not attempt to.

To raise the level of their thinking, it is important for people to raise the vibratory rate of their physical vehicle. Here again it cannot be done just with the mind. It is fusion of the mind and body that will raise the vibration, which is, in reality, the raising of the consciousness. It is a combined effort. You cannot sit in an easy chair and accomplish a true raising of the consciousness without physical activity to the extent of which you are capable.

If this physical activity is limited, the raising of the consciousness is limited. It really is a reciprocal system. What feeds into it is what can be garnered back out of it.

You can readily see that the education of a child as to the physical workings of the organs of the body is so important. This is where it all must start. This will lead to the ability to visualize, which will lead to the ability of the cells to understand their purpose and duty. This will in turn allow the mind to continue with higher visualization practices, which will in turn allow more intensity of action in the cells of the organs. This will again raise the level of intensity of the thoughts the mind can receive, to again transmit or broadcast to the entire physical system, the same as the large universe is constantly transmitting to all bodies in the universe.

The way this transmission takes place in the large universe is as follows: a huge electrical field is moving around the universe as the Milky Way galaxy through which our solar system has been passing for thousands of years. This field is what extended off of the first nuclear explosion. It contains the subatomic particles which were splatters of the first matter in all the universe – splatters of the condensation that formed in the vacuum. This field is made up of energy waves which are made up of thinking masses of subatomic particles.

These particles were encoded with electrical bits of energy which preceded the actual birth of the mind of the Creator. They were the substance of which the mind of the Creator was formed. As the subatomic particles drew together in a "like attracts like" manner, the mere fact of mass that grouped as positive, negative or neutral caused a sparking of ideas and thoughts. When these ideas and thoughts then met with opposition from the other electrically programmed subatomic particles it caused friction which, in turn, caused new ideas and thoughts to come into existence. This was the forming of the mind of the Creator.

Without friction there can be no growth. Friction is the instigator of growth. Here are two factors that can be directly applied to humans – the grouping together into like mind for mass electrical energy and the meeting, then, with opposing electrically programmed individuals to allow friction to take place.

This electrical field in the large universe is constantly transmitting thoughts through electrical impulses. Each planet or star or sun picks the thoughts up on their level of understanding. These electrical thoughts are in a graduated state of intensity. This isn't stated as being on higher levels of understanding for there are no such progressions in the cosmos. The progressions are made by intensity alone. When something is of higher vibratory rate it will receive higher intensity.

In a human being, fourth dimensional thought is a higher, finer electrical vibration than third. It does not consist just of the content of thought, but also with the electrical vibration that it rides on throughout the body. The higher vibratory wavelength is more powerful as far as influencing your body, all parts of your body, and is needed for the healing of your body.

Fourth dimensional thought will not consist of negativity. The minute you think a negative thought, it is so slow and rough a vibration that you will immediately drop back to third dimensional thought. Negativity cannot exist in the fourth dimension. The higher, finer vibratory rate cannot hold a negative thought. It is incapable of it. Higher, finer vibrations are always stronger than lower, rougher vibrations, even though it might seem to be the opposite. At first glance, something vibrating at a low, rough speed might seem more powerful, but the wear and tear on the instrument holding this type of motor breaks it down. Fourth dimensional thought patterns, being in harmony with the energy waves, will not dissipate and will not cause wear and tear on the body. Therefore, a person using them will appear to retain youthful capabilities much longer than one who is not. The secret to not aging is, indeed, in the thought processes.

The physical condition of your body when it sends out a thought is what determines the power of that thought. The drive and energy behind a thought is directly related to the drive and energy being put forth by you when you send out the thought in your daily functioning. Age would play a role, as would health play its part. Astrological signs, personality traits, psychic ability, mind control training, education, balance, the totally integrated person that your are is what determines the power your thoughts are sent out with. Of course, the emotion connected to the thought plays a large part and this emotion is made up of all the various aspects I just mentioned. The importance of a thought to one thinking it is what determines the power level used for third dimensional thought. When you reach fourth dimensional thought, it will be automatically stronger than third. The levels of power for fourth dimensional thought will be different between two people for all the same various reasons given above.

In a battle between third dimensional thought and fourth, fourth dimensional thought will eventually be the winner no matter how strong and powerful the third dimensional thought is. This is due to fourth dimensional thought's relationship to the energy waves it travels on. It will be consistent and ever present in the universe, whereas third dimensional thought dissipates over time and must be constantly sent. Fourth dimensional thought has long-lasting ability. Therefore, if you can do this kind of thinking, you must, indeed, watch your thoughts very closely. You can learn to do this and will find yourself able to go into "neutral" as far as thinking goes, at times. This is something that is not possible with third dimensional thought.

Going into "neutral" with the mind means that you strike a balance between third dimensional thought and fourth. It means you find a compromise between old, outdated attitudes and the new concepts that you either don't totally understand yet, or aren't ready to accept and incorporate into your being. It is usually the latter. These new concepts aren't comfortable and for that reason aren't something you can readily incorporate into your daily living. Therefore, you keep them in your mind, analyzing them, visualizing daily events that they could have been applied to and thinking through to the possible outcome had the new concepts been used.

In the meantime, you simply aren't reacting with the old attitudes anymore. You are in "neutral" as far as responses to situations that you are trying to change. This will enable you to adapt new ways of approaching situations. As far as what happens in the brain itself, it is in "neutral" gear. It takes a "wait and see" attitude, but the old response will have been shut down. This will have a tendency to look like you have closed your emotions off, but in reality you have not. They are in a state of transition.

Fourth dimensional thought is the thought level used by those who are considered brilliant or of genius capacity. It is a clear, unemotional level of conscious thought. There is no room in it for the lower emotions but there is wisdom and compassion and spiritual love. If you achieve this level of thought and wish to remain there you will no longer be as interested in earthly physical acts, including the sexual act. One must be ready to give up these pleasures to enter fourth dimensional thought processes. They will block your entry quicker than anything else. Too much focus on sexual activity in your physical life will block your entry into the fourth dimension. The fourth dimension will take place on the Earth in your physical body, but your physical body is only there to serve as it is needed to progress the work of the fourth dimension.

As we progress even further into the future, there will be no attachment to things of the physical body. This will include what you eat, what you wear, your sexual practices; in other words, all enjoyment of the physical body will be done in your mind/thought. You may experience the delicious taste of food in your mind while your physical body is eating something totally different. You may actually be eating something very nutritious to supply this fantastic mind with energy so it can continue to remain in the fourth dimension. This will be a long way into the future.

Multi-level beingness is the goal of the fourth dimension. Unfolding out of the chrysalis is the way to achieve communication with your inner being. This will, in turn, lead to multi-level thought processes that will allow permanent entrance into the fourth dimension, which is, as stated, all in the mind/thought.

The brain works as a computer in blocks that multiply as they are stacked. The mind works the same as a multi-level sales company if you want to think of this as an analogy. Someone locked into third dimensional thought does not have this multiplication factor influencing their thoughts. Present day schooling methods are not formatted into leading the thinking processes into multi-level thinking. This is where new teaching methods such as the Chrysalis Schools (to be described in a later book) are needed, to not only open the creative channel, but to also train the thought processes by visualization exercises to expand into many layers of thought at once. These schools will be based on integrating the arts into all subjects. Music, especially, is one of the best methods to bring about multi-level beingness.

The heart chakra, the fourth chakra of the human body, is the transformer from the third dimension to the fourth dimension. The lower three chakras have been almost totally opened and in use on most people. This is true more strongly in some than others. Very few individuals are using the higher three chakras: the throat, third eye and crown.

Each chakra consists of twelve levels of growth. When you are totally open all the way through the twelve levels of the third chakra, you are standing on the brink of the fourth dimension or the further opening of the heart chakra. Everyone's heart chakra is opened to the fourth level at birth. The pushing upward of the energy will begin to open the fifth level. This is the level of working with others, knowing that we are all one body of the Creator.

You may have chosen to work mostly through the heart chakra in a particular lifetime. If so, the other higher centers will open gradually, but your focus will remain in the heart chakra. This means you will be working in the area of helping others transform their lives into fourth dimensional thought, which is clear intelligent reasoning, and not remain stuck in the metaphysical field of the higher third dimension. People must go through this field but need to keep moving and not get stuck in psychic phenomena. For those of you who chose to work through the heart chakra, this is an area where you will be of much service to others – prodding them to keep them moving.

Many people believe that the further you go into psychic phenomena, the more spiritual you are and the more progress you are making. This is very far from the truth.

These people are getting stuck by communicating with deceased friends and relatives who were at the third level of the third dimension when they crossed over and know nothing about the fourth dimension and the sheer, glorious beauty of the mind when operating at this level. When in the fourth dimension all the bad attitudes of the past are gone and replaced with joyous, happy, intelligent and wise thoughts of realizing that you are not separate from others; that truly we are all part of the Creator.

Going deeper into meditation will tend to hold you caught in the third dimension also. You need to start using your animal-based brain, influenced by your higher mind, and keep yourself firmly on the physical Earth plane while living your life with intelligence and integrity, not going around in a daze with psychic happenings surrounding you. These psychic experiences will continue to happen and tend to hold you stuck if you don't wish to move on past them. You need to know that you must desire to move on past them and demand it of your mind to search for something that has more meaning than meditating and keeping yourself anesthetized against the problems of the real world. Fourth dimensional thinking is the only way these very real problems will ever be solved.

Fourth dimensional thinking will also have a natural tendency to clear up any problems of the physical body. One example of this would be to apply fourth dimensional thought to eating. This would be to cut your food intake in half by thinking that each bite is twice as big as it is. In the fourth dimension this thought will be so powerful that it will cause your hunger pangs to subside. It will make your body think that it has had enough food and your weight will immediately begin to slim down. This is not as difficult in the fourth dimension as it would be in the third. It is almost impossible in the third and this is why diets will never work. If in fourth dimensional thought, you can eat anything you desire by applying the thought process of doubling the amount. You will find yourself enjoying everything you like and eating less and becoming thinner. It can all be done with the thought that you are eating double the amount you really are. It has to be a very strong thought; so strong that you actually believe that the food expands. When you find yourself leaving amounts of food uneaten, this will be the signal to begin to take smaller portions.

Incidentally, this was the principle on which the feeding of the five thousand was done by Jesus. He was acting on the principle for others, not Himself. None of us are anywhere near being far enough along to be able to accomplish this for others, but we can accomplish it for ourselves. The fourth dimension is where miracles happen or how miracles happen. Those who achieve miracles are always living in the fourth dimension or higher, most of them higher. Fourth dimensional thinking is actually the beginning of miracle making.

Everyone should experiment with this concept of doubling their food. With so many people overweight at the present time, it could be the impetus to allow many people to strive to raise their consciousness to the fourth dimension. This is also the thinking that allows people to exist on breath alone. They are reaching very high mind levels when they do this and are so strong that their body thinks it has had a banquet.

The second example of fourth dimensional thought has to do with time. In the fourth dimension there really is no time such as third dimensional life is controlled by, but when living on the Earth you must operate in a time frame. That makes it difficult because, in reality, you will be moving very slowly in fourth dimensional thought while time is flying by on the Earth plane.

In the fourth dimension time is motionless. It does not go anywhere. It just is. When in fourth dimensional thought processes you could leave this sentence half finished and return to it a week later and it will not have moved. It will still be in the ethers, dangling just where you left it. Because of this, it is much easier to get stuck in the third dimensional passage of time. You see third dimensional time flying by and realize that you feel like you are standing still when in fourth dimensional thought. You feel as though you are being left behind and would rather be in the fast moving third dimensional time. This, then, causes you to slip back to the third dimension.

Think about an airplane way up in the sky. It looks motionless, does it not? You cannot see that it moves very fast. This is where fourth dimensional thought is in relationship to third dimensional thought being you on the Earth looking up at that airplane. In the fourth dimension everything stays pretty much at an even keel, but if flitting back and forth as you do in the third dimension, things get very confusing and frustrating. You feel like you are not making any progress at all. It seems as though you are regressing at times, but in reality you are growing and expanding inwardly even though the outward appearances would not seem to verify this.

Time in the fourth dimension is equal to molasses instead of sugar which is what third dimensional movement could be compared to. Please contemplate time and relate what I'm saying to your own life.

The third example of fourth dimensional thought has to do with being "present in the moment." To be "present in the moment" does not mean you forget the past. You merely know that it is dead and cannot be resurrected. You take the good lessons you learned from it and put them in your backpack and continue your journey. The unenlightening experiences were not wasted by any means. They allowed the good experiences to take root and grow and settle into your being.

If it were not for the unenlightening experiences – which, at the time they were taking place, your mind tried to tell you were wasting your precious time – you would not have recognized the good experiences. They taught you things you need to carry with you as part of your foundation.

Accept what each day brings and use it for personal joy and experience. What another person thinks or feels should have no bearing on what the day's experiences are bringing to you and what meaning they are giving your life. This is being "present in the moment." It has nothing to do with your relationship with other people but has everything to do with your relationship with yourself. It is a spot on the map or carpet of time/space that is you.

To be “present in the moment” is to feel exactly what is happening inside your body, the physical sensations, and correlating them to what you are assimilating through your mind. This must all take place simultaneously.

As far as your relationship with other people this is being “present in the universe.” It is a much bigger, wider, more all encompassing spot on the map. To be “present in the moment” may be compared to the head of a pin and “present in the universe” to an acre of land. Understanding this is most important.

To be “present in the universe” you have no thought for yourself but rather as to how your interaction is affecting those you are interacting with. For instance, a thought of how to apply one of the seven virtues, such as love, and how it would help someone else is directed by you to this other person. If it is to someone who needs to live his life showing more love for others, he will automatically start doing this.

Being “present in the moment” is self-growth on a personal level; being “present in the universe” is self-growth on a universal level. You need to do both. At any one particular time a certain incident may strike you as being either a personal level experience or a universal level experience. You must follow your gut reaction in this and determine which way you will use the experience. Both ways – either being “present in the moment” or being “present in the universe” can take place when alone or with others. It is the state of mind one is in that will determine which way the lesson will be used.

Many on Earth have gone past fourth dimensional consciousness into fifth dimensional and even higher consciousness. The number of people who have done this is so small when compared with the overall population, that it will not lift Earth into the fifth dimension. Actually, we are just at the point now where a glimmer of hope can be seen that Earth can be lifted into the fourth dimension in the area of space allotted to her in the universe for this purpose.

By that, I mean that Earth's life span is laid out as though on a piece of carpet running down a hallway. There are markings on this carpet as to where she should be when other universal events have moved into their new positions. When she comes to the end of the carpet she will return to the inner planes. If she has not attained her goals by the time she is rejuvenated and reborn up out of the black hole, she will review, as do humans. She will have to review for a certain period of time until reaching the plane she was at when she went into the black hole and continue from there.

Each time she is reborn she attains a higher degree of consciousness until she no longer has to reincarnate. At this point she will become a star. This is very, very far into the future for planet Earth. In fact, even reaching the fifth dimension is very far into the future for her in the way that we figure time. Earth cannot be of fifth or even fourth dimensional consciousness until her inhabitants are.

Part IV

After the Cataclysms

Chapter Ten:
Higher Vibrations

The Harmonic Convergence in August of 1987 actually began the transition period between the Piscean and Aquarian ages. This was the cosmic event that changed the picture of imminent cataclysms that were slated to happen. It was an astronomical cycle that was a medley of many cycles. The large cycles of the universe are on an astro-plane chart in the laboratories of planet ten, the experimental planet of our solar system. The cycles of lesser bodies are on individual charts also. The best way for us to understand their chart would be to emulate it by charting cycles on transparencies and overlaying them. If this were done, the triangular configuration that shows up just preceding the large cycles of approximately 26,000 years would become obvious. This is the energy that causes a leap in human consciousness. By doing the cycle charts, we would quickly discover this and many other puzzling things about the relationship between astronomy and humanness.

There were two triangle alignments that raised the planet

up and are even now holding her steady as she moves into her new position. One of these two alignments of stars and planets was in the shape of a triangle with the North Star at the apex of the triangle and the star, Norma, as the right hand of it and the star Alioth (or Saturn) as the left hand of the triangle.

The star Norma is only in position to affect the planet in the fall of each year. It can be found on August star maps. My birth in late September was planned carefully because this is when the star Norma is at the height of its influence. My name was chosen before entry: Norma for my home star, Joyce, my middle name, for the Spirit of Joy and Green, my maiden name, because the color of the fourth dimension is green. Green is also the color of the Earth and the color of the reasoning intelligent mind.

The path the stars move on is a flat parallel moving belt and it is moving in such a manner that at other times of the year the star Norma is in the furthest point of the belt from planet Earth. The star Norma was coming out of the esoteric mode at this time and was ready to assume its place as the right path of the three pointed alignment. It is this alignment of creativity with the Creator God that is sparking off the enormous amounts of energy at this time.

At the same time, Earth was in alignment with two other stars also forming a triangle. The point that Earth had at that time was the left-hand arm of her triangle and is on the side that places Earth and the star Norma side by side. As given, the star Norma had just been born into exoteric activity and was full of fresh new ideas and ways of expressing creativity.

The alignment that started in August 1987 will be in effect for several hundred years before bypassing each other and their influence on Earth. The leap in consciousness has already started and will continue to accelerate and be aided and abetted by those on the higher planets. They always work with natural rhythms. The cause and effect from this alignment will become obvious to all if they open their minds to possibilities beyond known knowledge.

As given, this alignment is sparking off enormous amounts of energy. Anyone who wishes to "tune in" to this new energy has only to desire to be in the creative stream of energy and show this desire. This can be done by making an attempt to bring forth some new, creative, imaginative idea or article that will give humanity hope and inspiration. This hope and inspiration can take the form of letting them know that there are better ways of doing everything connected with the exoteric life on planet Earth. The most applicable of these ideas will come to those who are tuning in through artistic, creative attempts to express their inner being. What a glorious opportunity for all! These ideas and concepts will be interpreted differently by each and every living thing on the planet for a wide variety of changes into beauty everywhere.

As usual these ideas and concepts will need to be stepped down to Earth plane interpretation. We are all in a position to do this. No idea will be wrong. No idea will be too far out. No idea will be laughed at. As a matter of fact, each "far out" idea will serve to stretch the mind of others. This, then, will allow an idea to enter that other mind and, perhaps, not be as "far out" as the original idea, but will incorporate that idea into something that can be used now.

The meaning of "drawing near to the fourth dimension" is the increase of vibrations that all are feeling. Many of you are nearing the vibratory rate of the areas you planned to plug into. These higher vibrations you are feeling are preplanned vibrations to attract you into your life's goal. As these vibrations increase, they are raising the planet's vibrations. It is not the vibrations of Earth causing the change in you. It is the change in your vibratory rate as you pursue your goals that is changing Earth's vibratory field. As everyone pursues their spiritual goals, the new vibrations are raising Earth up to become a spiritual planet.

It is not the new high vibrations that are causing your personal cataclysms. It is the old, outdated, low vibratory patterns that you are still stuck in that cause your personal cataclysms. Earth responds to those who are expressing through her, which all of us are. We are but a mobile extension of the stationary aspects of Earth. When the vibrations of a certain area are low and negative due to the attitudes of the people, the planet responds with physical activity. The low vibration actually shakes the foundation of the fine, sensitively balanced inner core of the planet. This is the same as low vibrations shaking the fine, sensitively tuned inner core of each of us and causing cataclysms in our lives.

This, then, will cause you to start thinking of preset goals and raising your vibration in order to find these goals. As these vibrations increase, Earth physically starts sending out the vibratory pattern that you programmed into it before entrance. When you have experienced a certain area of the Earth in a past incarnation, that area will glow when you pass over it in your thoughts. The areas you chose for this lifetime, and are expressing in and through, will light up the same as the lights on a switchboard. They will vibrate with a certain rate that you need to match in order to have all things available that you planned. Without this matching vibratory rate the doors will not open. The world for you is dead until you light it up with what you chose for this incarnation. It then lights up and starts sending out the pre-planned vibratory message. You have to hone in on this message to take advantage of all things that you set up prior to incarnating, to make this life work.

Being on Earth gives you an opportunity to take the material being offered by Earth and forming your own world according to your own perceptions. It is all in the mind. Earth is inactive until you bring to life for you the part of it that you want to experience. Say, for instance, you want to have an experience of sailboats. When it comes time for you to manifest into the physical, you will chart your path by plugging into the computer program that gives all the details about sailboats. You will plan your lifetime using these "pieces of the game board" to decide when and where you will be born. It will also help you decide what it is about sailboats that you really want to experience and how you want to use this experience to better the world. You will have much help in making these decisions, but at the same time you also will have much free will in designing your game plan.

I once had a small glimpse of the world as it is seen on the inner planes when one is planning their incarnation. The inner planes are, of course, inside the mind and I momentarily accessed an area of my mind that I had not reached before in this lifetime. This is why this glimpse has stayed with me. I now know that this area is available not only to me, but also to everyone on Earth. All you has to do is stretch and reach for it and the world is yours.

You should all be about this stretching and reaching. We're all too close to the goal now to give up. You will be complete master of your life and all that takes place in it. Do not compromise your beliefs any longer. Take advantage of any opportunity to open the door wider onto your possibilities and bring your world to life. Light it up and make it start vibrating and continue reaching for that vibration.

Many of you reading this were brought into manifestation during the time of the old energy. Therefore, your physical forms are not open to receiving this new energy without deliberately placing yourself in a state of openness and asking for the new energy to enter your body. It is not going to be automatically available to you. You must actually learn how to access it.

You will go through a period of preparation if you have opened yourself to the new thinking, the new consciousness. This period of preparation will last for a different length of time for each individual. In some it will last literally for years; in others it may take only six months. It will be a period of intense distress for most, not only physically but also emotionally. Some people started going through the preparation stage before the energy was even available on the Earth plane due to an inner awareness that there had to be something coming that would make all the agony worth while.

On the other hand, children will be very open to this new energy, but they must be educated about it. They can be helped, as far as accessing it, by means of musical vibrations. Believe it or not, this is what is behind the sudden interest in and popularity of the stringed instruments, especially the violin, at the present time. These youth seem to know instinctively that their body needs to have the vibration of the strings and by playing the violin it is physically closer to their brain and will affect the dissolution of the membranes more quickly than playing the piano, where the vibration is physically further away from the brain.

Until this new energy is totally available, people of all ages will suffer from more headaches than usual and more muscle aches and more broken bones, especially young people. The bone structure of youth is of such changed DNA that the vibratory field of Earth is causing a brittleness to the bones and a stiffness of their muscles. These youth need to trust their intuition as to what will help them.

This new energy also accounts for many turning to drug usage. They know they need to have something and think it may be the drugs that will help them. The only thing mini-cycles of extremely high frequencies that circle the solar system very fast. This energy will become stronger as the nearness of this cycle approaches. The cycle will actually "ride" with the Earth for a period of about five years of Earth time and the effect of this energy will recede for about ten years.

Most cycles that affect the planet have a period preceding them and a period of recession after them, but do not ride with the planet for any length of time. They are more like two ships passing each other. This particular cycle is like one ship actually connects to and sails with the other for a certain time period. This gives a much greater opportunity to take advantage of all that this cycle of the Creator's thoughts have to offer. With the other cycles it is like one ship disturbs the water for a certain period of time approaching and leaving. With this new cycle of energy it is as though the ship disturbs the water for a period of time as it approaches, keeps it disturbed for quite a period of time before leaving and also keeps it disturbed while leaving. In other words, the water is disturbed until the ship is clear out of sight.

Those who open to the new consciousness before it totally recedes will be the last to be able to join the Intergalactic Command. The Intergalactic Command is made up of all those in manifestation and out who are vibrating at the new rate. When this vibratory rate enters one's body it eventually opens the membranes between the layers of the brain. These layers were opened on me by having the vibration of music with me every day of my life from the age of four on. For those who did not have this opportunity, the new energy can have the same effect.

Most people will have to work very hard to take advantage of this cycle of energy. Those of you who are already into the new consciousness will need to stay attuned to it by meditation and vibrational tuning methods that will be all around you before very long. There will be much experimentation with the new vibrations taking place soon.

When the realignment of the planet is completed, the vibrations on the Earth plane will be much more intense than they are now. This will have the effect of speeding up time. Also, all of our machinery will run at a faster rate of speed and will run much smoother. Our music will sound much clearer; even people's speaking voices will sound different. Not only will sound be affected, but visual images will also become clearer, sharper and more colorful.

This is only a small part of the benefits to people that will come as Earth makes the move. This move is going to be very slow and controlled due to the help we are being given from those on higher planets who are able and willing to do this for us. Earth is very special to other planets in this galaxy. It is almost like the "spoiled darling" of all because of its uniqueness and also the fact that it has been an experiment, an experiment that benefits all others in the galaxy.

Some examples of what the change-over into the fourth dimension will be like are as follows: third dimensional science will become very dangerous in the fourth dimensional world. More and more "like attracts like" circumstances will take place causing destruction of a man-made nature to all sorts of situations. Third dimensional electricity will become very erratic in a fourth dimensional vibratory field. New methods of refining electricity to a higher transmission rate will be very necessary, not just a faster transmission rate, but a more highly refined current. Because of this, we will have to carve a path for new technological changes, but carve it we must or the world will slip back rather speedily to the old ways that must be outdated, as far as technology.

This new energy will actually allow the development of the new technology much more easily because the vibrations will be easy to access and harness into structured forms of technology. The time will approach quickly when new technological projects will be able to be attained without the cloudiness that surrounded them in the past due to unclear vibrations of the old energy. The concepts which seemed so new just a short while ago are much more able to be understood in this new time period. For the five years that the new energy is riding with us, there will be a continuous spread of new technology over the whole world.

This will be the start of the new culture, or new age that has been predicted for so long. This is the goal, the new culture, the new energy that will encircle the Earth and lift her high to take her place in the shape of the universe. As the move is made, some of the other benefits to people will be as follows: not only will the clearer, cleaner air make audible sounds sharper, clearer and louder, but at the same time the higher vibrancy of the energy itself is going to have an effect on people's ability to hear. The ear chakra will begin opening more as the move is made and the new vibrations start working on the physical form. This, added to the ability of sound to travel more easily is going to change the broadcast frequencies of our radio, TV, telephone and anything that is produced as audible sound. Our recording technology is going to have need of changing and updating the way it is done.

New products using voice recognition will be much easier to program to get a clear sound. In fact, this technology will be very popular as the move of the Earth is made. It will eventually be combined with a robot thinking ability and serve as a companion to the blind, the elderly, the lonely and the ill, replacing the need for someone to constantly watch over them. The robot will be able to be trained to be almost human. It will have the ability to not only communicate but to also sense the feelings of the one it is communicating with. It won't be long after that until the use of cameras will be added to the robot giving it a visual image to go with the data that it is sensing in other ways. Then the robot will have the ability to take on even more responsibility for those who need to have constant care.

This is only one of the new things that will be a result of the move of the planet. Another would be in the growing of food and of all plants and flowers. They will need to be monitored as to getting too much sun as the move is made. A needed technology will be brought into being using a filter between the plants and the sun. Large sheets of filtering material will need to be stretched over fields to prevent them from being harmed by ever-increasingly more dangerous rays coming from the sun. The filtering material needs to be of an organic kind, not anything plastic because this would have a disastrous effect on the plants and those who eat them. Also, cotton will not work as well as the new organic filtering material, but will work much better than plastic, especially in the beginning. Plastic simply must be done away with as far as being used outdoors and should be totally done away with eventually. The more intense rays of the sun will cause a transmutation of the make-up of anything plastic that will cause it to be unpredictable and dangerous, especially any plastic used in transportation monitoring equipment.

We are even now having problems with some of our modes of transportation on the Earth. They are being affected ever so slightly now in the beginning of the move of the planet and will become increasingly so as time goes on. Airplane problems are due in part to the slight movement the planet is making. This has the tendency to throw off machinery on the Earth that is controlling our travel, whether it be airplanes or trains – anything that is mechanically controlled. Even radar is being affected.

Those in the transportation industry really need to be aware of what is taking place and have a research team working on reprogramming equipment at all times. This will require constant adjustments to sensing equipment. Transportation equipment will need to be adjusted because transmissions will be ahead of themselves. They will travel ever so slightly faster. This will not cause big problems right away, but the point will come when their speed of travel will be just enough faster than the clocks that are controlling them that it will begin to really have an effect on transportation.

Our animals will be affected, as well as us, in many ways other than just their hearing and seeing abilities. Their taste buds will be affected, as will ours. Their smelling ability will become sharper, as will ours. All of this will have its own effect on the physical form. Just the ability to smell other things more sharply and clearly will affect our sinus cavities which will in turn affect our eyes. In the beginning this will be a problem. There will be much allergic reaction to smells that did not bother us previously. We will have more headaches also, due to this sharper smelling, hearing and seeing ability.

There are many changes coming and everything in our society will have to be constantly updated. Research on all areas must be ongoing to keep up with the change.

If the shift of the planet had happened as fast as it was first feared, it would have been too fast for most of us and our society to survive. As it is now, it is going to be hard on the elderly and the ill. The children will adapt and adjust much more quickly than their parents. All of the new technology given to those working on a better future needs to be perfected as soon as possible. This will help those in the higher planes who are using their energy to help our planet make a slower move. Those of us on planet Earth simply need to be aware and to do our part to make the move easier for everyone.

Our politicians need to become aware of what is going on with our planet and make provisions to change laws more quickly to keep pace with the changes. This is the only way to prevent complete chaos. For instance, when new energy becomes available through the efforts of researchers, there must be a way for it to be taken advantage of as quickly as it becomes available. This must be done without fear of reprisal from existing electric providers. The same thing holds true for all new technological advances. They are going to start coming hot and heavy. The pendulous system of our governing bodies will be the undoing of our social structure and also the safety of all the people in the world.

Our government must become enlightened. This is the only way we can survive this immense change that is overtaking our planet. The time is now for everyone to become alerted as to what is taking place. Those in the higher realms will help slow the process of our move as much as they are capable of. At the same time they need to have the cooperation of those of us on the Earth plane itself to do what we can to help ourselves.

Most people are at the brink of taking the first step into the fourth dimension. It is part of the 11/11 energy spiral that took place. The vortex of energy surrounding the planet, caused by the friction of unpervaded space meeting this spinning ball, is stepping up as the planet moves into uncharted virgin fields of space. This vortex is, of course, causing an increase in the spinning of the chakras in each human body. Those of you who are not "opened up" to the knowledge of metaphysics will notice it as physical illness. Those of you who are open to higher knowledge need to realize what is happening and adjust your thinking to more focus on spiritual thoughts and less on the physical. You need to have less concern about what you eat, how you dress and how much enjoyment your physical body can have. You should turn your thoughts instead to what is coming with the fourth dimension – the new thought patterns, etc.

As people "open up" to assimilate the more powerful vibrations, they raise their own vibrations to their higher etheric selves. This allows contact with the real fourth dimension on the etheric level of the planet. It retains its beauty and color and high vibratory rate as they bring it into existence on the Earth plane – the exoteric plane. The unmanifested plane is the esoteric (spiritual); the manifested plane is the exoteric (physical), but it is all just a difference in vibratory rate.

The fourth dimension on Earth's ethereal body is shining clean and sparkling new. As given, there are a few already there who have progressed from third dimensional reality but, for the most part, the beings there are from the higher planets. They are reaching back to get the fourth dimension ready for the masses of humanity to enter when they are ready. This is when the Christ will reappear and not until then. He is waiting.

Chapter Eleven:
Life in the 4th Dimension

There is an area of confusion as to the fourth dimension being on Earth. In reality it is both on Earth and also has a mirrored reflection in the realms around the Earth. It is on Earth now in our minds/thoughts. The unmanifested fourth dimension is, at times, intermingled with the manifested fourth dimension in our minds. Sometimes it is up above slightly, but at any particular given time the part of the Earth's surface that is in the full moon stage is nearer the fourth dimension, the unmanifested Earth's surface, the Earth's ethereal body. Planets have ethereal bodies as do humans, and just as physical healing has to take place first on a human's ethereal body, so also does a planet's healing have to take place on the planet's ethereal body first.

Those on Earth experience the fourth dimension in their minds; those on the astral planes experience it as reality. It is the same vibratory rate and they can be in contact with each other if they so wished. Perhaps those on the astral planes see something from their overview that would benefit those on Earth. If this is the case, they can communicate it at night when the person on the Earth sleeps. At this time, he goes to what is called the inner planes, which are part of the astral planes. The inner planes can also refer to the higher planets and even the astral planes around them. There are strictly divided divisions on the inner planes just as there are between the big planets, but those on Earth can reach them all depending on their mind dimension.

When space ships leave the Earth's atmosphere they dematerialize. This is because they are from the other dimensions, perhaps the fourth, fifth or even higher. It is a matter of higher vibrations. It is not the same as seeing ghosts or apparitions because they are third dimensional. It is the next step up in evolution and very few are ready to enter it.

The fourth dimension is a thought-form, an ethereal, unmanifested dimension of Earth. At the same time, the fourth dimension is a matter of a level of thinking and a way of behaving that comes from those in the fourth dimensional mind measurement. You might call the fourth dimension the "research and development" department of planet Earth. Theories are tested there first before being allowed to drop down and cast their shadow on the third dimension. For example, our new technology is seen and thought of and planned all in our minds for quite a while before it is brought into manifestation on the Earth plane. All future developments come from the higher planes and are dropped down. At each stage they are translated into meaning for that particular plane.

The new culture being formed across the world has

already cast its shadow. You see, this new culture is already in place and operating smoothly in the fourth dimension; it just needs to work down from there. The patterning is done on the fourth dimension and worked and reworked to be sure it operates smoothly and then is sent down.

When you enter the fourth dimension, your physical body will no longer be in control of your spirit. In third dimensional life your physical body, including your animal-based brain, is in almost total control. When entering the fourth dimension of the mind, your spirit controls your body. This is the ultimate goal for the lessons of planet Earth.

Some of you are intuitively tuning into realms on the upper left side of the brain which would be the 7th, 8th and 9th dimensions – the use of the reasoning intelligent mind. What you need to have at this time is a creative outlet for the right side. This will mean dropping back to the 4th, 5th and 6th levels which are in the right side of the brain. For some of you it's as though your progress went straight up the left side of your brain, progressing from 1st, 2nd and 3rd to 7th, 8th and 9th dimensions, making you a real left-brained individual.

I would encourage you to try writing poetry or learning music or perhaps trying to draw. Any of this will help integrate more spiritual connotations into the intuitive "tapping in" that you are doing. The balance is not there as far as left brain, right brain. Perhaps you sense the spiritual aspects and only need to have a little exoteric practice in the arts to bring the esoteric in much more strongly than it has been.

As your mind works mostly on day to day situations, you are drawing almost constantly from one area, the ninth level. If you could cross over into the right side and eventually tap into the tenth level, the tenth planet of experimentation, then more lucid, new ideas would join with the intelligent ideas already there. The way to do this is, as I said, drop back into the 4-5-6 levels, a stage you most likely skipped. This would be elementary drawing, elementary music, poetry, whatever it takes to start opening that door.

Perhaps you have been working on achieving fourth dimensional thought for many lifetimes and have finally had the breakthrough. Then this is the lifetime for you to learn wisdom, which is the correct use of fourth dimensional thought processes. This will be the only chance you will have to gain this wisdom in an earthly experience. When you cross over you will be able to apply it only in thought, not in physical action. Then when you reenter in the next life you will have the wisdom but will not have the experience of incorporating it into everyday life.

The next lifetime, then, will be mostly used up learning to use this wisdom in your daily life. You can see that if you do not learn to use wisdom in this life, you will set your progress back another lifetime. All of your other aspects that are not in manifestation at this time need for you to learn it now. You have the honor of being the one manifesting in the outer world and achieving the wisdom of the ages. Now your other aspects want you to continue on and learn to incorporate it in your daily life. This can't be stressed enough. If you have the wisdom of the ages available to you but you cannot function with it wisely with your fellow man, then your life will be miserable and your achievements a waste.

The first area you need to incorporate it into is in your relationship with others, especially with your children, your spouse and the rest of your family. Physical families are the first area that needs to be focused on when making the transition to the fourth dimension. This is because any leftover karma with them must be cleaned up or you will not be free to progress on in the true fourth dimension. This leftover business will continually pull you back to the third dimension. If you have achieved the breakthrough, you must start to apply the wisdom of incorporating fourth dimensional thinking and actions. I urge all of you to do this as soon as possible.

Once you have opened this fourth dimension there is no turning back to third dimensional consciousness. The wisdom will stay with you; the use of the reasoning intelligent mind will follow you for the rest of this incarnation. It has been asleep in most of humanity until the present time.

Most people have not yet reached the breakthrough into the wisdom of the ages. Some are close, but until they make the total breakthrough they cannot live fully in the fourth dimension. Those who have made the breakthrough need to understand this about them. It will help you be able to accept them as they are and this is the first priority for those living fourth dimensional life. On the other hand, when those in the third dimension make the final breakthrough into unlimited knowledge they will understand you better. They also will, at that time, be able to relinquish the pleasures of the flesh.

As far as looking into the future, much of it is already set. It will come about because you planned it that way, not only in this present life, but before you entered and also during your previous life. Therefore, you need to greet each day with joy and curiosity about what it will bring and whether or not it is bringing you what you previously determined you wanted.

Fourth dimensional living will take care of the problems of earning a living naturally because all of this is relegated into the higher thought processes where it flows and happens without any forcing. It truly is worth giving up third dimensional longings and cravings because even your spiritual life takes care of itself.

In other words, all areas of your life will begin to flow much more smoothly, as though on automatic pilot. This will leave you time to apply much of the knowledge gained by the wisdom of the ages. You will be able to think a thought and know where to apply it. Once you know what situation to apply it to it will automatically be applied with little or no effort. Your original thought will guide it to its destination.

Living in the fourth dimension is very powerful. Your thought has the power to change the lives of others, not just your own. You had to change your own thinking to reach fourth dimension. Now you must use your thought to empower others to change their thinking and be able to achieve the fourth dimension themselves. They, in turn, will help others by applying their powerful thoughts to changing others. It is like one grain of sand at a time being transformed into glass and turning back and selecting several other grains of sand to help. It is a long process to change over most of humanity like this, but it can be done.

Peace on the Earth plane can only come when it is finally realized that the answers for each individual or nation must come from within themselves. Peace, not contentment or full joy and happiness, but at least the prelude to these goals will be accomplished. More must be awakened to the fact that answers must come from within. Any method that can bring realization of just this one fact is very necessary at this time period because things are critical on the Earth plane today. People need to understand about third dimensional life and what can be done to bring it into the fourth dimension.

The meaning of the ethereal realm or fourth dimension is that of a future state that is present in thought in the current or "now" time period. This is what all higher dimensions are – the future that is present in thought and can be contacted and then used as a goal and the means to reach that goal.

The overall goal of the fourth dimension is to attain joy in living each day to its fullest. This is not necessarily living in the moment. It is a balanced plan of action drawing together the physical, emotional, mental and spiritual aspects of yourself into a whole person. You will become a self-expressing, loving, caring, intelligent and compassionate person. You will interact with all others and be responsible for your own well-being and financial support and mental attitudes. This is not going to be easy due to your emotional body make up coupled with misinterpreted teachings and outdated social systems. I'm not giving this to be discouraging, just to make you aware of the complexity of the goal.

Not too many have seen the overall goal. It is incomprehensible to someone who has not progressed through at least the ninth level of third dimensional growth. A few have seen the goal and every time they reach out to others with a hint of what this is they receive looks of shock and disbelief. This is because the fourth dimension is that of separation. The sparks were together on the third dimension and will be again in the fifth, but the path of the fourth dimension will always be that of separation. This will be true no matter on which level you are viewing it: macro, human or micro. This is almost a prerequisite of the fourth dimension, the ability to function separately from others. Perhaps this will help explain why there is so much separation and divorce.

As we go much further into the future, there will still be marriages, but mostly between those still in the third dimensional processes and the teachings of the religious institutions. For those truly in the fourth dimension there may be a coming together long enough to give birth to children, but these parents simply will not feel the need to live together. The children will not suffer from this. They will be raised by single parents until they are old enough to attend child care centers.

The child care centers in the future will be very much improved over today's child care methods. They will be structured to allow the children to become much stronger individuals. They will enter these child care centers as soon as body functions are under control. Here they will all receive huge doses, equal doses, of love and attention. The people in charge of the centers will necessarily be very advanced souls. Male and female alike will share these roles. There will be no more separation of roles according to sex. This has been very harmful to humanity in the past and still is a huge problem. In the future your sex simply won't matter.

It is towards these children that the educational guidelines in the Chrysalis Teachings are directed. But until the parents are ready to enter the fourth dimension of the mind it would be useless to try and instill these guidelines of education into any schools. This holds true whether the schools are being run by religious institutions or those professing to be "new thought" schools.

The schools in the future will be true fourth dimensional, educational Halls of Learning, Halls of Knowledge and Halls of Wisdom. These schools will lay out a path for achieving fourth dimensional thought. The first, the Halls of Learning will be based on the awareness stage. The next, the Halls of Knowledge will be for applying the learning to day to day living. This is the stage of implementation. The Halls of Wisdom will be in preparation for fifth dimensional thought processes.

Sometimes those on Earth can visit the fourth dimension

during their time on the inner planes at night while sleeping and bring back a vision of what life is like in the fourth dimension. At one point I had an opportunity to do this. I had a vision that was a visit in my mind as to what a city on Earth will look like in the future after all have made the transcendence to fourth dimensional thought. This will be after the forming and changing (done by cataclysms) takes place, which is Earth moving into a new orbit. This move will make it closer to the sun and moon both. It will also make it closer to the other planets. It is moving into its true orbit due to another planet moving into the solar system which is causing Earth to move a slight bit closer to the sun. Magnetic attraction has captured this new planet and as it settles into an orbit around the sun, all the other planets will shift slightly. It doesn't take much of a shift towards the sun to make it just enough more dangerous that all will have to live undercover. Also, the shift will make it closer to the moon, which, because it is dead with nothing to attract it closer to the sun, will stay in its same position and we will be much closer to it. It will light up the night sky to the point of daylight when in the full moon stage.

The fourth dimensional city that I visited was the home of my guide. He shares this city with others of like mind. To be of like mind they must be of the same mental level of thinking because form wouldn't be real to one from a higher realm. The vision was a view of future existence of life on Earth with much new technology in use. This view of the fourth dimension is still in the mental stages at the moment but getting ready to begin manifesting on the physical planet.

The vision I saw of this city is not so far off, less than fifty years, most likely, until Earth has made her move closer to the sun. At the time of the visit I was still in the higher levels of third dimensional thought but was able to visit it because I was so near in thought to those who had just crossed into the fourth dimensional thought on the astral planes. Their vibrations were able to be averaged out into a blending that allowed us to be in communication with each other, although I was not able to be in communication with any in the city. I saw them, but could not communicate with them. Nor could my guide at this point as he had lowered his vibration slightly. We were just onlookers and could see the people, but I could experience the food and plants and animals because they are not as high a vibration as the people were. The following is what I experienced when I visited the fourth dimensional city through the vision.

When I visited the city, it felt empty, but it wasn't. We, my teacher and I, went at a time when all were sleeping. There is not much difference in day and night due to the artificial atmosphere. As we entered the city one of the first observations, I made was that everything seemed to be pinkish-golden. Also, everything seemed to be undercover. There were lights on inside some of the homes and yet, out away from overhead protection, the sun was shining in the park.

The parks had large fountains and pavilions of colored panels with the sun shining through like stained glass allowing one to bask under whatever color he chose. There were bright colored birds in the parks, tame birds like pets.

The next thing to strike me was how clean everything was. The houses were dome-like structures made out of a white substance that looked similar to concrete but wasn't. The homes could be added to with other domes as needed or desired. One dome was a swimming pool, another a health room, etc. The walls in the health room had variegated tubes of colored lights around them. The swimming pool rooms in the homes were filled with golden light and much greenery. The bathrooms were filled with mist and greenery.

The streets were made of the same substance as the dome houses. They were lined with plants and trees. There were tiny open aqueducts above the sidewalks carrying water to the trees and plants and also putting moisture into the air. There were even dog toilets placed along the streets, and dogs were trained to use them. The street lights were glowing poles.

They had food prepared for my visit. I was served a meal in a restaurant, consisting of meat that was a cross between chicken and rabbit. Plums and pears that were cross-bred were sliced on the plate with the meat. There were potatoes that had been crossed with beets; they had red centers when sliced. There was also a salad with large leaves that I understood were grown under water. There was a seaweed dish similar to spinach and the bread was dark, rough and delicious.

There was an airfield with individual airships, something going round on a pole winding them up like rubber bands for a power source. These were used for short trips.

Bathroom waste went into large underground vats and was converted into gas for neon lights that were used in growing plants and also to light the homes.

Indoor orchards were in all stages of growth at the same time in different areas. The ceilings in the fruit and vegetable barns were panels able to be any color of light, depending on the different fruit and their different stages of growth. The attendant could tell if a tree or plant was not doing well by feeling the vibrations upon entering the room.

The same was true of the musical motors running all systems. One could sense harmony in the vibrations if all was well, and could pick out the one motor that wasn't functioning right. These vibrations would sound a harmonic chord internally in one's body if all was well. There were open cars that could hold two persons. These cars could be folded flat and stored in racks like bike racks and were plugged into a power source while people shopped.

The department store I was taken to had a large domed ceiling filled with greenery. There was a greenish cast inside all buildings due to plants. In the department store, one could sit and view the merchandise passing by on a big screen. If you saw an item of clothing you liked, you would be shown how you would look in it.

The schoolroom I visited had tiny open aqueducts to water the plants and put moisture in the rooms.

My overall impression of the visit was of the beauty and

cleanliness of everything.

This visit to a fourth dimensional city took place in my mind through soul travel to the outer atmosphere surrounding the Earth. The soul can fly like this and circumstances were arranged so that all seemed real to me. On the fourth plane there are "things" that are as real to them as our things are to us, such as furniture. It is quite difficult to understand because where we are now is not reality. It is an illusion, but yet our Earth is real to us. Our walls and furniture are real to us but not to anyone from the fourth dimension. They do not exist for them because their higher vibration makes it possible for them to walk through our walls. One coming from the fifth dimension to the fourth would not find form real on the fourth dimension.

It is all a matter of vibrational levels. Anyone coming from a lower vibrational level as I did and visiting the fourth dimension would feel their walls and ground. They seemed to have form the same as on the Earth. It is not physical matter form though, it is ethereal matter form and those inhabiting this city were in ethereal bodies. Most of them will soon be back on Earth or are already back and share this view of a future city with me. They experienced it fully; I only experienced it as though seeing a movie.

Sometimes a movie you see here on the Earth plane is so real you feel you are there and sharing the experiences, even emotionally, with those on the screen. This is the way I visited and experienced the city – as though watching a movie. When seeing a movie you may feel what the characters are feeling and sometimes you can even imagine the taste of what they are eating, etc. But you cannot communicate with them in any way and you cannot change anything about the scenes you are seeing. Each dimension exists at exactly the same time. Normally, one would have to pass over to the inner planes of the third dimension to enter the fourth. But during the crucial time period when it was thought that there were going to be violent cataclysms as Earth made her shift, there were those who entered through being lifted off by space ships. These ones who were lifted off from the planet as experiments to see if the process was workable are in the fourth dimension now. They are all well and happy due to special care and education. They were all offered the opportunity to return to the third dimension and a few elected to do so, but most chose to remain. Those who did return had vague, dream-like memories of where they were.

The unmanifested realm of the Earth includes all twelve dimensions of the inner planes. Those on the inner planes have lives and experiences within the mental realm that are ongoing at all times. They manifest on whatever dimension they are located on but never in physical manifestation. Nevertheless, they are on the Earth plane atmospheric conditions.

The new times are here. What are you waiting for? Are you going to join in the glorious days ahead, or are you going to be left behind? The choice is yours and in the long run who will miss you if you do not come along? You are only holding back your larger self, your higher, more spiritual part, if you do not come along. Those in the higher planes hold out their hands in brotherhood to all no matter what your past has held, no matter how small your belief system. Now is the time, the grace period or the amnesty period, if you will. Just keep moving forward in all areas of your life. God is action, remember! Will the God Within you join with the God Within those of us who are moving onward and upward? I sincerely hope it will be so.

Chapter Twelve: Coping with Change

Weather changes are being caused by friction in Earth's outer atmosphere. Not only is this friction being caused by the planet's move into a new orbit in virgin space, a natural occurrence, but our space stations will also cause friction which then causes weather disturbances. They present a blockage of the flow of the jet stream of air currents. This will happen with each space station put into orbit. Each one will have a bearing on the weather in a different locale depending on the altitude and size of the space station.

If our scientists would but think for a minute what a circling fly or mosquito does as far as agitating a human, they would get an idea of what their manmade satellites are doing to the planet. I know you're all familiar with those big flies that buzz very loudly and never stop flitting around. They just don't sit down anywhere or if they do they're only there for a second. At night they are drawn to lights and buzz around the light bulb. Our man-made satellites have the same effect on the planet. We have much to learn about what can and cannot be done as far as interference in the atmosphere surrounding our planet.

These space stations need to be outside the Van Allen belt to keep the weather disturbances at a minimum. The cause of the recent severe drought in Africa can be directly attributed to the orbiting satellite that our country sent up. The drought will become increasingly severe until a breakthrough as to cleansing the atmosphere takes place that will allow a natural moisturizing and cleansing of the land.

Perhaps some scientist will read this and understand what is being given and have some sensible guidelines drawn up. These guidelines should be for what will or will not affect the planet and to what degree it will affect it, by comparing it to the human body about which they have this information. When more interest is shown in this area, there will be more available. For the time being there needs to be some thought going in this direction. Think, people, think about how we're treating our planet. As an example of this, about twenty years ago I was moving a TV set by myself and it started to fall. I caught the full weight of it on my right wrist. I went to a doctor who x-rayed it, said the bone was cracked and put it in a cast. Within a short period of time, perhaps two weeks, I became so nervous I couldn't stand it and went to a different doctor. He instructed me to take the cast off immediately. Having the cast on had caused the nerve and tendon to grow together. Because my fingers were not in the cast, every time I moved them or tried to use them, it affected the nerve all the way up the arm and into my neck. The cast had been on long enough that I had to have surgery to separate the nerve from the tendon. By this time my whole body was being affected. This can be compared to what takes place on the planet. It will affect the whole planet, not just the immediate area.

The planet is a living entity and has a "nervous system" as such. Its nervous system is composed of topographical phenomena and data such as rivers and mountain ranges, etc. When a mountain is disturbed, believe it or not, the nervous system of the whole planet is disturbed. When a river is dammed up, it is as though stopping the flow of energy to certain nerve trunk lines of the planet.

There is no way to separate the components that make up our planet. These components are the land and water masses, both on the surface and underground, and the atmosphere around the planet. Everything done to the surface of the planet affects the underground part and everything done underground affects the surface. Everything done in the ocean or the sky affects not only the ocean or sky, but also the land masses.

Because its natural resources are being used up at a rate that is not being replenished, the planet cannot continue to renew itself, as far as life cycles go. They cannot be replenished in time to prevent a barren surface within approximately 50 to 75 years if the present rate of consumption is continued.

There has been a set of circumstances started that is similar to a chain reaction in physics. It will not be too long before it is irreversible, but as of now it is still reversible. This will cause loss of vegetation which will cause loss of both animal life and human life. We could end up with these two kingdoms wiped out. This would put us back to a planet with no life forms in manifestation for the souls to enter to experience and gain knowledge to help them in their evolution. This would set back all of the universe. This is not meant as a scare tactic to frighten anyone, just that this can and will happen if the stripping of natural resources continues at the rate it is presently being done.

Even now the forests are becoming sparse due to a drop off in natural reforestation. The changes will be gradual, but some thought should be put into the situation now instead of waiting until the forests are depleted any further. Preliminary plans should be made to save them at the present time. Several different species of trees in the northern part of our country should be re-seeded in other areas in the south to save the species. The elm, the oak and most hardwood trees are in danger. These can be grown in climate controlled nurseries to begin their growth. They can then be transplanted back into the northern areas of the country after a period of five years and they will then survive.

The climatic changes that harm them are periods of drought that will not allow the roots of the trees to receive the necessary moisture. This is due to many reasons, one of which is a distortion in the natural availability of reserve underground water supplies. These underground streams should be mapped and what is happening will readily be understood. At the same time that these are drying up, the natural amount of rainfall will be decreasing in the northern areas of our country and increasing in the southern areas. This will be due to the new positioning of the planet, the new alignment that she will be going through. It has actually already started. It will be a gradual realignment, but will cause many changes in climate on our planet.

Because our air will become ever increasingly cleaner, our water supplies on the Earth will also change. There will be a more equal distribution as the move takes place and all the moisture will be cleaner. As the Earth becomes cleaner, the atmosphere will become cleaner and will send back cleaner moisture. It is a reciprocating process.

The reason the air will become cleaner is due to several things. One of these will be the fact that as Earth makes the move into its new orbit, it will leave behind the cloud of pollution around it. This cloud of pollution will have nothing in it to attract it closer to the sun, the same way our moon will not make the move. Therefore, we will actually move away from this pollution band and leave it behind. When this happens and our planet is totally removed from this band of pollution, the pollution will be cleaned up by other planets with means available to them. We will then be left without this protective band that is filtering the sun's rays for us.

This is what we are seeing by the tear in the ozone layer at the south pole. It is the beginning of the move. This is a natural happening and is partly causing the "greenhouse effect" or global warming, but it is a natural happening.

Actually, there is a twofold cause for the hole in the ozone layer. One: we are beginning to make the move away from the polluting band of atmosphere. Two: it is being caused partly by elements on the Earth attracting some of the pollution and "eating it," you might say.

The first cause, the moving away from the band of pollution, has already been discussed. To explain the second cause, the band of pollution around the Earth is being caused by chemical combinations from industrialized nations. Those chemicals left in the natural state do not cause pollution, only when combinations are made that are at disharmony with the mineral kingdom. Chlorine is the one chemical that will separate itself from any combination and return to the pure state. If it has been combined with only one other chemical, then that chemical will also return to the pure state and not cause pollution. If it has been combined with two or more chemicals then they must be tested as to harmony after the withdrawal of the chlorine.

Concentrations of chlorine are being attracted to the south pole by the presence of arginine found there. It is acting as a magnet and holding the chlorine close to the Earth instead of allowing it to dissipate. This is helping to cause the hole in the ozone layer. This is because the concentrations of chlorine will continue to build higher in the atmosphere, due to the magnetic pull of the arginine, instead of spreading out lower to the Earth. These concentrations of chlorine will not spread out because the arginine itself is concentrated in one particular spot. To disperse the arginine, it would need to be mined and "ground up" due to its being a solid metal in form, and then distributed equally over the surface of the Earth. The chlorine would then separate out of the pollution in equal amounts over the Earth's surface and not be concentrated enough to cause any problem in the ozone layer.

Part of the pollution is being attracted to the Earth and being cleaned, even as this is being written. Other elements found in the soil are also attracting ingredients out of the band of pollution and absorbing and eliminating them. It is not possible to do this over the whole surface of the planet, because these elements are not present in the soil in all other areas. If they were, then we could clean up the band of pollution, but even if we could clean it up we are still going to make the move away from it.

The proper starting place for correction would be in stopping the pollutants that are going into the air. Control of emissions from manufacturing plants could be done through applying the Science of Music as the criteria as to what is safe and what is harmful to be dispersed into the air. The Science of Music is what those from the higher planets will use to clean up the pollution band after Earth has moved away from it. They will use the process of "like attracts like" as found in the Vibratory Charts given in the Science of Music. (These will be discussed in later books.) Those chemical combinations whose vibrations form harmonious chords that are part of the Harmony of the Spheres will not harm our atmosphere. These would be the combinations that fall into perfect mathematical categories laid out by the theory of music. There are many possible combinations that will actually cleanse the air and clean up the pollutants already there. Much research needs to be done in this area.

Climate changes will take place everywhere. The change in weather patterns will continue to be felt and weather will not stay as it has been anywhere. It cannot stay as it has been; it will be ever changing. The area I'm in, western Colorado, will become very wet and stormy. It is going to be safe, but after the repositioning of the Earth it will not lend itself to easy living. It will be a difficult area in which to raise food.

The central part of the country will be the most livable area and eventually many people will end up there. Missouri and surrounding areas will be ideal for crop growing as they are now, but even more so afterwards. Areas that were tropical will continue to cool off and other areas that were hard hit with moisture will slowly begin to become drier. It is all going to take place over a rather long period, extending past the lifetime of most of us on the planet now.

The state of Florida will not have any more instances of land dropping into sinkholes than what could be considered normal activity for the state. In other words, the state will not have an earthquake and will not be underwater. There will be a few isolated cases of land falling into sinkholes, but this has been happening forever in this area and is certainly nothing new.

If anything, the state of Florida will find itself entering a rejuvenating stage, but a transitory stage. It will be reclaiming land lost to swamps and sinkholes. It will actually become a much more stable land mass than it has been in the past because the climate of the state is in a state of change. The climate will change to a less tropical climate. It will change to a dryer but warmer one in the lower area and a dryer but colder one in the northern part of the state. At that time it will no longer be as conducive to citrus fruit growing. It will become more suited to range land for animals. There could be a huge game preserve planned for the reclaimed swampland. This would include birds and aquatic life. This is how safe the state of Florida is going to be.

There was a time period when the picture did not look this encouraging, but as the universe is ever changing, the dismal picture seen a few years ago of the coast lines of our whole country being under water has changed.

As far as more earthquakes in California, at one time it was thought that the "big one" wouldn't be for twenty years. This is now debatable. This depends on what happens. That was the time frame that was seen for it before things got so bad. It could be speeded up; it could be slowed down. It's up to all of us. By raising our consciousness from third dimension to fourth, we could eliminate all of these catastrophes, but that probably won't happen.

Nostradamus predicted an earthquake in California for the time period of May 10, 1988. Many things have changed, due to the free will of humans, since Nostradamus saw the earthquake in this time period. For one thing, the pressures had been relieved by venting that was done in other locations. The time table for a major earthquake in California will be when all other volcanic and earthquake activity has dwindled down to where there has been no major venting for a period of three years. The next major earthquake will extend the full length of the state, breaking off the land on the west side of the San Andreas fault. When Nostradamus saw this happening it was on the map at the time period of 1988, but other cataclysms, including the eruption of Mt. St. Helens, were not on the map and this activity has rearranged the map that Nostradamus read.

If these guidelines are followed, that of no major venting for a period of three years, then the "big one" can be predicted fairly accurately as to time. People can be evacuated and little loss of life will occur, but of course much loss of property will take place. It would be inadvisable to invest in any property west of the San Andreas fault line.

Earth is and always will be undergoing formative processes, the same as each individual life is constantly undergoing change until the withdrawal of spirit. Until that time one must live one's life without fear of catastrophe. If you are in an area where you are constantly afraid of cataclysms happening, such as earthquakes in California, then I would advise you to move to an area where you will not have this constant fear. Otherwise, accept what comes in the way of Earth changes and the changing climate conditions as a challenge, an exciting challenge, to pit your skills against nature. This is what can lead to new inventions for combating severe weather and also new ways of predicting what is going to happen. So, live your life and enjoy it to the fullest without fear; with joy and appreciation for the opportunities presented to you each and every day.

Where would humanity be if it had constantly worried about doing anything because of the possibility that there might be a cataclysm a week, month or year down the road? The constant talk of cataclysmic upheaval is only adding to the upheaval in people's lives.

Some of the predictions of cataclysms will come true simply because of the mathematical chances involved in guesswork. This is what most of the predictions must be considered: guesswork. Even those entities on the unmanifested plane of Earth do not know when any particular area is going to experience a cataclysm unless they study it scientifically. They do monitor certain areas, but there are not enough of them to monitor every square mile of Earth's surface both underwater and on the ground. Many of the cataclysms manifested on the land have their start underwater.

We must realize that what some people are seeing and picking up on as far as cataclysmic activity is a reading of past records, not future records. It is as though they are reading outdated newspapers. This should help alleviate some of the consternation these reports have brought to people.

Be about it! Erase the word cataclysm from your vocabulary. Replace it with upheavals, upsets, storms or release of tension. Storms are very necessary to clean out the old and replace with new thought on an updated evolutionary cycle.

As stated before, the preparations that everyone should make are to work on attitudes that will hold them back from entering the fourth dimension. People need to concentrate on those attitudes that are not completely transmuted. This will prepare them for the change over to fourth dimensional consciousness. Earth herself is entering into a period of rejuvenation. Are we, its inhabitants? Most of us are lagging behind.

It is a releasing of the tensions of birth, of the elements that brought about the solidifying of matter, that cause what is termed as cataclysms. There are certain areas on the planet more predisposed than others as to releasing these tensions. Some of these are known and simply should not be inhabited. Move to where you have not this fear and then release it and live your life as normally as you can, to achieve the goals you wish to achieve, in the area you determined to achieve them, if possible. One cannot be given a yes or no on whether to live in a certain area. You must learn to listen to your inner being. You must make your own reality.

There always has been and always will be areas that are uninhabitable for people. These areas will continue to change. We have evidence of tribes of Indians moving on from certain areas in the past and we surmise that it was weather conditions that caused them to do so. This is correct and will continue to be what must take place. It usually doesn't happen now because of advanced techniques of combating adverse conditions.

All that has happened in the past will most likely happen again in approximately the same general areas because these are "weak" spots in Earth's surface that readily allow for the release of tension. These spots ought to be revered as sacred because without them there would be no Earth for us to inhabit. They are very necessary and cataclysms, by rights, should be celebrated events because they allow the continuation of life on planet Earth. Be not afraid of cataclysms. Welcome them instead as they are cleansing processes.

The Earth changes that are coming are good changes. All change is good. God is action, remember, and this applies to the planet herself. Lack of action signifies withdrawal of spirit. Knowing that the planet is alive with spirit adds a zest and excitement to your life which the constant dread of cataclysms certainly isn't doing. Excitement and wonder and awe of the Creator should replace the dread.

The new culture that the change over into fourth dimensional consciousness will bring about, is not going to be ushered in with big cataclysmic events. That was the picture at the birth of Earth, not her middle age. Middle age does denote changes though, the same for planets as for humans. There will be adjustments in the life force more suitable for advancing years; more graceful movements in the life force, more gentle emotional upheavals – a steady, more balanced physical nature.

Yes, Earth is approaching middle age. Compare this to natural changes that take place in the human female.

When outside events surrounding the woman are inharmonious so also, then, do the internal changes become more violent and stressful. When the middle-aged woman has her life in balance emotionally, she goes through the period of menopause with very little upheaval; so, also, the Earth. Ponder on this and live your life without undue concern about approaching doom. It is not going to happen that way. The cataclysmic events have ever been with us and ever will be. It is an ongoing process of life.

Earth will not go through major disruptions again until she starts the downhill side of her journey and that is a long way into the future. At the present time we are heading into what is being called the "golden age" of Earth. This would compare to the active adult years for a human after the turmoil of adolescence is over.

The turmoil of adolescence made up the cataclysms of the planet that happened earlier. The new cataclysms will compare to the stage of menopause in the human female when cataclysmic events again take place. So it will be with Earth. As with humans, there will still be upheavals and dark periods in their lives when in their active adult years, but nothing so cataclysmic as the teenage years. Therefore, there will still be cataclysms and weather changes throughout Earth's journey, but the uphill and downhill sides of this journey are where the devastating cataclysms will be felt.

The cataclysms are, indeed, going to continue but they will be stretched out over a long period of time. When Earth begins the downhill side of her life, her liquid center will grow calm and cease to cause disturbances on the surface. Compare this to humans who achieve the same status as they grow older and have started on the track of closing down operations. Only with Earth itself the time period is expanded to such great lengths that the comparison cannot truly be seen by many.

Those entities on other planets can see over the hill and into the future. They know what a long process it is going to be to enlighten humanity and prepare it for the next period in the 2500 year cycle when the Christ Consciousness will be more readily available, more easily available to all. It will be halfway into the new culture before the time comes for the return of the Christ Consciousness; five hundred years into the "golden age," which we are on the cusp of now. This is the change over period. The new age has begun, the old age has not left. This is the mixture.

These cusps last probably 25 years on each side of an age which gives us a 50-year cusp period. The more awakened, the quicker the transitional period is. This is the purpose of the Chrysalis Teachings, to ease and speed up the process of the cusp period between the Piscean Age and the Aquarian Age. The new age has started. The old one has been fading away probably since 1965 (beginning of cusp period). This means the year 2015 will probably be when an exact date could be put on the Aquarian Age as saying this is it now! In the picture of it appearing and moving across a screen, this will probably be approximately when the Piscean Age has disappeared totally off of the screen with the Aquarian Age in full view. As it is now, both can be seen on the screen.

The volcanoes and earthquakes that are taking place now are all part of the realignment of Earth in virgin space. Once she is settled into her new area all will return to normal. This will take a period of approximately twenty to thirty years. She will be mostly into the new space in about ten years but the settling into the new orbit will take longer.

We are now in our year 1997, so there is approximately 19 years before the new culture is all there is. The old will be gone. This is a very short time to prepare people; to bring them up to where they should be at that time. Once that time gets here, it will be, as given, 500 years before the actual return of the Christ. He cannot be drawn into Earth's auric field, until the cycle is correct. He cannot return until the auric field of the Earth is cleansed and changed over from a third dimensional vibratory rate to a fourth dimensional rate. The auric field, the ethereal body around the Earth, must have the higher vibratory rate and the only way this can take place is through the thoughts and consciousness of humans. The consciousness must rise in each, to cleanse the thoughts that go out from each and contaminate the auric field around our Earth with lower vibratory frequencies. This latest repositioning of the planet is going to take place. The signs are all there for it to happen but the date cannot be accurately set. The best thing that everyone can do is try to raise their consciousness from the third dimension to the fourth dimension.

Life in the fourth dimension is ethereal – unbelievable joys and happiness for all souls who have suffered through the agonies and torment of the third dimension.

They have certainly earned this respite. Then the fourth dimension must progress onto the fifth. This is the closing in of all – the drawing back together of all creation, like the breathing in of a giant lung before expelling the next breath on the fifth dimensional level of evolution.

The oversoul itself is subject to this process of breathing that is part of the function of the central core where it all began. The central core itself is motionless, but above and below is churning energy which is being controlled by this breathing process, a huge pair of bellows, if you will.

Physical evolution follows esoteric or inner evolution. As the oversoul split and separated the sparks into individual consciousness, so also did the land mass of our planet split apart and separate into individual continents. As the oversoul will gather all individual consciousness back up to move on, so also will all the individual continents come back together and form into one land mass, ready to be used for the next holy purpose, after the rejuvenation of the planet on the inner planes, the black hole, to which she will return at that time.

I'd like to give a long range look at the oversoul's use of planet Earth and the exact process of cataclysms in connection with this use.

In the beginning Earth had a circular land mass. The land masses on Earth will eventually form back into this circle. The cataclysms are actually the "seven seals" that are misinterpreted in the Book of Revelations in the

Bible. The seven seals were the planned breaking up of the circular land mass into seven different areas. These areas were for the seven different sons of our Creator God Michael to use to begin their artificial insemination experiments on animals to bring forth the human being. These areas were where the seven different areas of creation were to start all at the same time. Two of the areas did not develop as planned due to two brothers (sons) not cooperating.

The seven original continents were America, Australia, Africa, Antarctica, Asia, Europe and Persia. Africa and Antarctica were the two where the beginning of human beings didn't start at the same time as the rest. Because the experiments in these two areas did not take place when they should have, they had to start life forms over again at a much later time. This was supposed to have been an automatic process encoded into the cells of planet Earth before birth. The other areas had already had five rounds of artificial insemination experiments before these two had any, hence they are behind both in divine evolution and physical evolution.

The Sumerians were the original colonizers of the planet. It was the Creator God Himself (Michael) who came to Earth at that time. He had seven sons whom the seven land divisions were for. As given, two of the sons did not cooperate and two of these areas were not peopled with humans in the beginning.

These Sons, Gods if you will, did keep coming and going but in the beginning they were the ones who artificially inseminated the animals with their genes. The land was still in a circle, all one land mass at that time. It was planned in a circular ring to start impregnation.

The seals were as follows: as each of the areas started developing human beings, it was broken off from the main land to fend for itself. The cataclysms – seals – were what broke the land apart. This happened at the time of the impregnation experiments because the force of the visiting space ships on the fragile thin crust of the Earth at that time caused earthquakes and volcanic action that caused each section of land to crack and break away. Africa was one of the last to break away. Antarctica never did and remains attached to the original inner circle of land.

The cycles of evolution that were talked about in part two of this book, will come around again in just the order the cataclysms happened the first time. They have continued to happen and will continue to happen. The breaking up of the land exoterically will not continue happening, but the breaking up esoterically will.

It is very difficult to explain this. Exoteric events and esoteric happenings will continue to cycle, both on a higher cycle of evolution. As the planet cooled, the cycle of earthquakes, caused by the vibrations of the space ships, were entered into the ethers surrounding the planet. When the earthquakes cycled back around they were weakened and also the planet was more solid, more hardened. This is why Antarctica did not break free. It was the last to be experimented on and Earth had cooled enough that the vibration of the ships did not cause the land mass to crack. Our space flights could conceivably

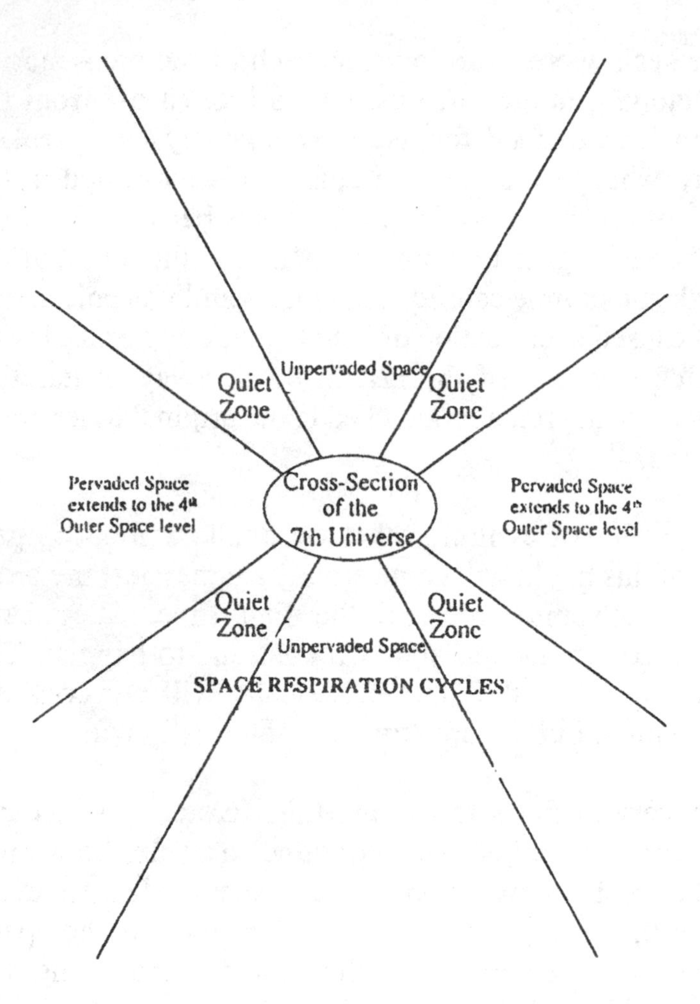

Space respiration cycles showing the quiet areas between pervaded and unpervaded space. This is the area, actually just a small corner of this area, that is moved into each time a solar system moves. — Illustration by Neil Hickox

cause the same thing to happen if a ship makes contact with a newly forming, cooling, body. Think of this.

Iceland holds the important position of protector of the sacred land called Greenland. Those in Iceland realize, more than most on Earth, the potential of not only Greenland, but also Iceland as being a treasure trove of undiscovered goodness to help humanity. Both of these lands will have their treasures unlocked when Earth climate changes so dictate. They will both become very important, but Greenland is the final location of all the continents pulling back in together to become the same as when it all started.

Perhaps this will help enlighten people as to the process they and their planet are going through. It should certainly help many in seeing a larger view and stretching their minds to contemplate what we are doing wrong. Think of the whole process as a very difficult video game and all the right decisions have to be made or the goal is never accomplished. Once it is understood, then progress in evolution will begin to take place more rapidly and can move onto the next stage of the Creator's evolutionary cycle. All are subject to this rhythm.

Our universe is not really going straight up or down, it is a flat-parallel moving belt around the central core as are all universes. Energy is above and below feeding life which evolves horizontally – in other words, out to the side. This fact should be studied more closely by our scientists in view of our planned space flights with an eye towards understanding better the cycles involved in all processes of life on all planets.

Earth's movements at this time are into a space that has never had any planet in it. This is due to the fact that not only is our planet moving through the cosmos but so also is our whole solar system ever so slightly moving at all times. It just happens that with each solar system there will be one particular planet that will orbit into unpervaded space each time. Because of this fact, even though other planets precede Earth in moving their orbits, the fact that the whole solar system is moving has allowed a space to be reached by our particular planet that has never known the intrusion of the solid body of a planet before.

When this is fully reached by the solar system and our planet at the same time, the upheavals on our planet will be stilled. This is because any moves from that time on will not be into virgin space, but previously occupied space.

There will be twelve solar systems in each galaxy and twelve galaxies in each universe. There is a mathematical breakdown of the move of these galaxies, solar systems, and planets that will see all of this quiet zone space pervaded by the time the twelve galaxies are formed. This will make up the seventh universe. At this time, there will be no more quiet zone and it will be time for the eighth universe to begin. It is all a matter of precise timing and perfect movement that allows each solar system to use its allotted part of the quiet zone to increase its vibration and rise up, taking its place in the scheme of things. This will allow the next solar system to use the area of the cosmos already pervaded by previous solar systems.

This previous pervasion makes the growth of the new solar system easier until it also reaches the point, the timing, when it must move and inhabit its allotted piece of the quiet zone as the instigator of higher vibrations. The same thing is true in the human kingdom. Those paving the way into the fourth dimension of thought will make the path easier for those who come after. But each individual person will reach that one area or spot that they must traverse themselves. This is an area of their own mind that they have not touched before. This opening is responsible for the personal upheavals that each person will face on the spiritual path.

In the large universe the whole system is precisely timed and planned as to insure that all available space is in use before the next universe begins working its way towards using all space allotted to it. These areas of space are building on top of each other. The unpervaded space above and below each universe is not used; it is the quiet zones between the pervaded and unpervaded space that are eaten up by the movement of the solar systems. There is still a "cushion" of unpervaded space above and below each universe.

This is difficult for us to understand. There is no way that our scientists can prove or disprove what is being given here as they have not the instrumentation to investigate this. When the Science of Music is furthered, this will be the beginning of understanding the movement of bodies of matter through the cosmos. Until then, scientists will have to accept these theories on faith and use them to try to improve their picture, their understanding, of what is taking place in the large universe.

The large universe is still expanding. Our planet is also still expanding out but almost at the point where it will contract and come together. This will be the drawing back together in the declining or contracting period of our planet. We are almost at the halfway mark in our expansion and the stationary period at the end of the expansion of our planet will be the "Golden Age."

Yes, the masses of land will draw back into their original circle pattern and form once again into the original circular home of the Creator God Michael. When this happens, He will be whole again (all people will have reached fourth dimensional consciousness) as also the land masses of the Earth will be whole. He will be ready to move onto the fifth planet where He will continue as a whole deity and all will be together again.

The process of forming back into a circular land mass must take place in the reverse order that it was split apart. As the exoteric pulling apart took place first and then the esoteric, so must the esoteric gathering back together take place before the exoteric can. This will be the land masses coming back together. When this all takes place, Earth will have completed her synthesization process of combining or "raising up" her 3rd dimensional activity to that of 4th dimensional activity for a smooth, harmonious 1000 years of no turmoil – the calm before the move onward begins.

About the Author and the Chrysalis Teachings

Norma Hickox is a professional musician who plays five instruments and composes and teaches music. She has been an organist and choir director for many different denominations and has also taught vocal and instrumental music in private elementary and preschools.

Her career has included painting, dance and theater, which led to her writing two musicals and to the recording of duo-piano music from these plays. She continues to write popular music.

When Norma began composing in 1981, the music and lyrics would flow out in complete form. A year later, the Chrysalis Teachings began flowing in the same way that the music had. The teachings are based upon the Science of Music, which is the fundamental vibrational order and creative nature of the universe.

The Science of Music gives an absolutely fantastic method of tying all kingdoms together in a grid of vibratory rates. These Vibratory Charts will lead to breakthroughs not only in the New Science, but also in agriculture, healing, pollution control, nutrition, the DNA code and space travel, to name a few.

The teachings act as a thought provoker, giving ideas that stretch our minds and break us out of the shells of old thinking, allowing us to wonder, "What if?"

The Chrysalis Teachings are in the realm of creative ideas, before inventions are made or new products are designed. They open the possibilities of new ways of doing things; of looking in other places or directions for answers to old and new questions. The higher plane concepts will help people understand and solve the spiritual and material problems of the Earth plane. This understanding will then assist the growth of the world into the peace and joy of the fourth dimension.

INFORMATION FROM THE AUTHOR

Updated Chrysalis Teachings are available on an ongoing basis in a bi-monthly newsletter entitled "The Chrysalis Teachings." One year (6 issues) is $25.00 in US currency made payable to:

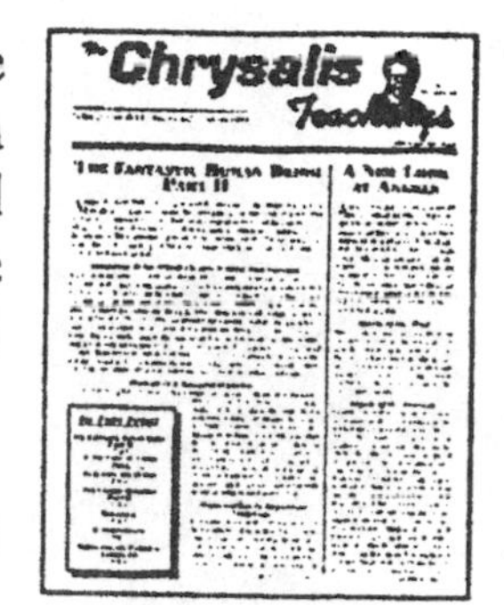

Chrysalis Publications, PO Box 3937, Grand Junction, CO 81502

Blue Star Productions is not responsible for the delivery or content of the information or materials provided by the author. The reader should address any questions to the author at the above address.